COMPLEX SYSTEMS, ARTIFICIAL INTELLIGENCE, AND EMERGING TECHNOLOGIES

Innovation in the 21st Century

Complex Systems, Artificial Intelligence, and Emerging Technologies

Series Editor: Klaus Mainzer *(Technical University of Munich, Germany)*

Published

Complex Systems, Artificial Intelligence,
and Emerging Technologies – Volume 1

COMPLEX SYSTEMS, ARTIFICIAL INTELLIGENCE, AND EMERGING TECHNOLOGIES

Innovation in the 21st Century

Klaus Mainzer

Technical University of Munich, Germany

World Scientific

NEW JERSEY · LONDON · SINGAPORE · BEIJING · SHANGHAI · HONG KONG · TAIPEI · CHENNAI · TOKYO

Published by

World Scientific Publishing Co. Pte. Ltd.

5 Toh Tuck Link, Singapore 596224

USA office: 27 Warren Street, Suite 401-402, Hackensack, NJ 07601

UK office: 57 Shelton Street, Covent Garden, London WC2H 9HE

Library of Congress Cataloging-in-Publication Data
Names: Mainzer, Klaus author
Title: Complex systems, artificial intelligence, and emerging technologies :
    innovation in the 21st century / Klaus Mainzer, Technical University of Munich, Germany.
Description: New Jersey : World Scientific, [2026] | Series: Complex systems, artificial intelligence,
    and emerging technologies ; volume 1 | Includes bibliographical references and index.
Identifiers: LCCN 2025026627 | ISBN 9789819813131 hardcover |
    ISBN 9789819813148 ebook for institutions | ISBN 9789819813155 ebook for individuals
Subjects: LCSH: Technological innovations | Artificial intelligence | Dynamics
Classification: LCC T173.8 .M34 2026
LC record available at https://lccn.loc.gov/2025026627

British Library Cataloguing-in-Publication Data
A catalogue record for this book is available from the British Library.

For any available supplementary material, please visit
https://www.worldscientific.com/worldscibooks/10.1142/14312#t=suppl

Desk Editors: Eshak Nabi Akbar Ali/Rok Ting Tan

Typeset by Stallion Press
Email: enquiries@stallionpress.com

Preface

This interdisciplinary book series explores the dynamic intersection of complex systems, artificial intelligence (AI) and emerging technologies. Huge and permanently increasing sets of data (Big Data) in nature and society can only be handled by algorithms of AI, which request innovation of more efficient computer technologies. But AI is not any technology. In the meantime, the most valuable companies in the world are AI companies. Recent AI innovations demonstrate that they are the game changer of the world economy and financial world. But it is not only a question of scaling big money to become a game changer in AI. Recent success (e.g. DeepSeek) demonstrates that fresh ideas are decisive. Thus, above all, it needs investment in the resource of human spirit. At the same time, the living conditions and the limits of natural resources on this planet must be taken into account.

Therefore, it needs a platform for innovative research and critical discussions that bridge theoretical frameworks with practical applications across diverse fields. The series welcomes contributions that investigate the foundations, methodologies and implications of complex adaptive systems and intelligent technologies. It aims to foster dialogue between scientific, technological and philosophical perspectives, emphasizing how emerging innovations shape and are shaped by societal and philosophical contexts.

According to several prominent authors at the end of the last century, a main part of 21st century science will be on complex dynamical systems. The intuitive idea has been that global patterns and structures emerge from locally interacting elements like atoms in laser beams, molecules in chemical reactions, proteins in cells, cells in organs, neurons in brains, agents in markets, etc. by self-organization [1]. But what is the cause of self-organization? Complexity phenomena have been reported from many disciplines (e.g. biology, chemistry, ecology, physics, sociology, economy, etc.) and analysed from various perspectives, such as Schrödinger's order from disorder [2], Prigogine's dissipative structure [3], Haken's synergetics [4], and Langton's edge of chaos [5]. But concepts of complexity are often based on examples or metaphors only. We argue for a mathematically precise and rigorous formalism. In this framework, we define local activity as the cause of complexity which can be tested in an explicit and constructive manner.

Boltzmann's struggle to understand the physical principles distinguishing between living and non-living matter, Schrödinger's negative entropy in metabolisms, Turing's basis of morphogenesis [6], Prigogine's intuition of the instability of the homogeneous, and Haken's synergetics are in fact all direct manifestations of a fundamental principle of local activity. It can be considered the complement of the second law of thermodynamics explaining the emergence of order from disorder instead of disorder from order, in a quantitative way, at least for reaction–diffusion systems.

The principle of local activity is precisely the missing concept to explain the emergence of complex patterns in a homogeneous medium. This principle can also be defined in technology with, e.g. non-linear electronic circuits, such as Chua's memristors [7]. It is not restricted to a certain domain, but can be generalized and proven for the class of non-linear reaction–diffusion systems in physics, chemistry, biology, brain research, and technology. The principle of local activity is the cause of symmetry breaking in homogeneous media.

The local activity principle inspires not only models of the brain but also brain-orientated (neuromorphic) architectures of computers. It turns out that neuromorphic computing approximates the energetic efficiency of human brains and avoids the enormous increase in energy consumption with traditional digital computing. Obviously, traditional digitalization is closely connected with one of the most challenging problems of mankind on this planet: the increasing demand for energy with all its consequences for environmental and climate problems. Thus, the local activity principle strongly supports the request for sustainable computing.

In the age of globalization, the Earth system grows together with human civilization. The climate and ecological systems can no longer be separated from one another, but depend on industrial growth and energy policies. Global communication networks and infrastructures as well as financial dependencies of banks and states are driven by non-linear dynamics of complex systems. One of the main insights of non-linear dynamics is the emergence of systemic risks which are caused by the interactions of many factors and players in the whole system. The emergence of systemic risks from complex systems dynamics is a challenge for control tasks in engineering sciences as well as regulation and governance in social systems. We need early warning systems in the technical and natural sciences as well as economics and politics. Therefore, we consider applications of cross-over disciplines in economic, financial and social systems with the emergence of equilibrium states, symmetry breaking at critical points of phase transitions and risky action at the edge of chaos. In any case, the driving causes of symmetry breaking and the emergence of complexity are locally active elements, cells, units, or agents in dynamical systems.

Thus, the demand for local activity in societal systems leads us to a plea for local active ('entrepreneurial') agents. Entrepreneurial activities are not restricted to industry, but mean any kind of technical, economic and social engagement leading to new innovations. With respect to sustainability, we need innovations considering benefits for the whole Earth system, not only narrow-minded profits. In short, sustainable innovations must be the goal of entrepreneurial creativity. Anyway, according to the mathematics and philosophy of local activity, the interaction of agents in complex systems is not sufficient for innovations. Innovative creativity of local individuals is the driving force of entrepreneurship. But, of course, local activity of entrepreneurship needs interaction with other agents in order to achieve benefits for the whole system. Obviously, the principle of local activity involves deep challenges of ethics and responsibility in society. They will also be considered in this book and end up with a clear societal message:

Act locally and think globally with responsibility to the whole Earth system.

The key to a successful strategy for complex systems with AI is innovation from emerging technologies. People tend to prefer not to leave familiar and previously successful development paths than to focus on new innovations. Parts of the automotive industry would rather follow the 'path dependency' of profits in polluting diesel technology than switch to sustainable electromobility in good time.

However, every shareholder knows that you should never rely on a single share class, no matter how tempting it may be in the short term. A good portfolio is characterised by diversity and variety in order to do well in the long term despite all the crises.

The thesis is therefore that the pressure of the current political, economic and military crises should be used to accelerate the switch to sustainable innovations in order to move away from path dependency on fossil fuels. The difficulty lies in the fact that the Green Deal called for by the UN must be in line with national economies and people's living conditions and must not overburden them (e.g. green steel and green hydrogen). This is the only way to solve the real major problem facing humanity: the global environmental and climate crisis. It is therefore important not to focus on a single solution, but to bundle the entire technological potential in an innovation portfolio.

Like shares, future technologies are also bets on the future. Diversity also ensures resilient behaviour in an innovation portfolio in order to be able to react flexibly to the risks of the future and recover overall in the event of selective setbacks. Energy supply chains are just one example of the complex networks of modern civilisation, whose huge volumes of data and information cannot be managed without digitalisation and AI.

However, digitalisation requires a huge amount of energy, which is different for different computer technologies. As with the energy issue, digitalisation must not be based on a single solution, but rather the entire technological potential must be bundled in an innovation portfolio. In this book and this book series, classic digitalisation and AI are examined together with future technologies, such as neuromorphic computing (based on the energy-saving mode of natural brains) as well as quantum computers, quantum communication and quantum technology. They are realised on different hardware of quantum, neuromorphic or classical digital chips.

A supercomputing centre of the future will integrate all these different hardware realisations in a common platform for different purposes and applications. It will be used with high-level hardware-independent programming languages, and the user need not have any idea of the basic complex technology such as in conventional classical digital computing.

Recent breakthroughs of generative AI such as ChatGPT and DeepSeek play a central role in the global innovation strategies and will be highlighted in this book. The advantages and disadvantages of digital and analogue technologies must be weighed up against each other and combined in a 'hybrid' IT and AI so that this innovation portfolio, together with the energy issue, is also geared towards a sustainable future.

An innovation portfolio is made up of basic, bridging and future technologies that change, are abandoned and need to be replaced by new ones in the short and medium term. An innovation portfolio is therefore dynamic and must be constantly shaped. Methodologically, the mathematical theory of complex systems and non-linear dynamics can be used to model complex systems and networks in nature, business and society. On this basis, chaos and risks are assessed in early warning systems and translated into strategic action.

Against this backdrop, it becomes glaringly clear what has been seriously lacking in the recent past: strategic thinking. The era of visual flight and short-term political and economic interests led to disaster. What can we do on the part of science, research, and education? Answer: Learn to think strategically because only on this basis is responsible leadership possible. This is also a central concern of this book series.

Recent political, economic and military crises also show that strategic thinking and action cannot be limited to individual areas of research, technology and business. Innovation strategies are part of a global confrontation between the world's political systems. Innovation systems, according to another thesis of this book and the book series, are an expression of different value systems and interests of old and newly emerging centres of power. In the sense of complex systems, a dynamic equilibrium must ultimately be created that resiliently balances out shifts and changes in order to avoid global destabilisation and wars.

Ultimately, this book is about global innovation and leadership in order to prepare this planet for its global challenges. The last chapter is dedicated to the planetary exploration of the Universe. This is not science fiction, but it will be a destination of this species in the next step of global innovation from a scientific, technological and economic point of view. In the last century, it was not only the goal of the Moon flight itself but the breakthroughs of emerging technologies and innovation dynamics on the way to the Moon that inspired human creativity. However, as long as we keep pushing each other back in bickering and wars, there is a real danger that the red lines set by nature will be suppressed and crossed. But then it will be irreversibly too late as the mathematical models of complex system dynamics prove.

For me personally, this book and this book series document the development of my scientific work from complex systems to AI which are now connected. The research in complex dynamical systems was originally deeply rooted in physics. Therefore, it is not surprising that the Nobel Prize in Physics 2024 was awarded to two AI researchers. It is this kind of scientific transformation that opens avenues to innovative breakthroughs for mankind.

The book is written as an introduction to the topic of the book series in an informal style and can be understood without special knowledge of complex systems, AI, and technology. There are only a few sections with mathematical formulas which are indicated by a star (*). They can be skipped without losing the overall understanding.

Klaus Mainzer
Munich
March 2025

References

[1] Mainzer, K. (1994), *Thinking in Complexity. The Complex Dynamics of Matter, Mind, and Mankind*, 1st edn., Springer: New York, 5th extended edn. 2007.
[2] Schrödinger, E. (1948), *What Is Life? The Physical Aspect of the Living Cell Mind and Matter*, Cambridge University Press: Cambridge.
[3] Prigogine, I. (1980), *From Being to Becoming*, Freeman: San Francisco.

[4] Haken, H. (1983), *Synergetics. An Introduction*, 3rd edn., Springer: New York.

[5] Langton, C. G. (1990), Computation at the edge of chaos. Phase transitions and emergent computation, *Physica D*, 42, 12–37.

[6] Turing, A. M. (1952), The chemical basis of morphogenesis, *Philosophical Transactions of the Royal Society of London. Series B*, 237, 37–72.

[7] Chua, L. O. (1971), Memristor — the missing circuit element, *IEEE Transactions on Circuit Theory*, 18, 507–519.

Contents

Chapter 1

Introduction

In mathematical models of complex dynamic systems, a phase transition occurs at a critical point when the system state instantly changes to another [1] (Chapter 2). Examples from physics include complex molecular systems such as liquids, which instantly change into solids such as ice crystals when cooled at critical temperatures or into vapour, gas and finally plasma when heated. Mathematical climate models with millions of parameters change irreversibly into chaotic states at critical control values. Complex ecological models predict how biodiversity will disappear at critical control values. Global financial systems react to the smallest fluctuations according to the mathematics of complex systems. In urban systems, the transition from order to chaos can be triggered by localised disruptions to the infrastructure.

Conflicts and wars can also be modelled as complex dynamic systems. None other than Carl von Clausewitz laid the foundations for this in his classic *On War* (1832) [2]. Clausewitz thus shows how the dynamics of a war are determined by culmination points, in which the attacker's forces and energy are used up and flag and the dynamics can change as the defender switches to attack. Clausewitz's culmination points are reminiscent of attractors in complex dynamic systems, towards which currents and trajectories converge. In Chapter 7 of his first book, Clausewitz deals with the 'friction of war' and thus anticipates remarkable insights into the theory of dynamic systems and chaos theory.

'Friction' in physics corresponds to 'dissipation' and the increase in 'entropy' in reality, with which, for example, the oscillations of a

1

pendulum eventually come to a standstill. Clausewitz refers to the reality of war in which friction is a fundamental disturbance for the ideal strategic planning of a general's staff. The demand for the primacy of politics and the idea of a people's war catapulted the book *On War* to become one of the most influential works of the 19th and 20th centuries, which is not only still taught at military academies today but is also of fundamental importance for management and corporate leadership.

Are there early warning systems to predict the sudden change of system states? The trends of complex systems cannot be clearly calculated from planetary positions like a solar eclipse, but critical control values signal a phase transition with varying degrees of probability. The limits of calculability depend on the degree of complexity of the system models. Complexity refers not only to the number of system elements but also to the type and variety of their interactions, which lead to difficult-to-solve non-linear equations in the model. In this sense, ecological and economic models are highly complex compared to low-dimensional systems of physics, not to mention political models.

In retrospect, all political, economic and military development trajectories seem to be heading towards a present point in time. Corresponding observational data and analyses seem to support this in retrospect. But in many places, they contradict our own economic interests, political convenience and wishful thinking and therefore ultimately lead to incorrect assessments. A fatal and dangerous mixture for analysing and evaluating complex system dynamics!

In the recent past, the prosperity of a country and the functioning of entire sectors of the economy depended on comparatively cheap fossil fuels and raw materials. This example again reveals remarkable insights from the theory of complex dynamic systems. In the 1980s, the American economist Brian Arthur, following the ideas of Joseph Schumpeter, spoke of the path dependency of economic developments [3]. In complex dynamic systems, development trends are described as 'paths' in which individual development trajectories are bundled. In companies and markets, positive feedback effects can reinforce the chosen path. Smaller influences are pushed back and it becomes increasingly difficult to correct the direction of the development paths. In the end, a path is adhered to even if an alternative would have been better. Teaching processes become increasingly unlikely and can lead an initially successful development path to ruin.

One vivid example was the German automotive industry, which was in danger of missing out on the leap into electromobility. The reason for this was the extremely successful development path of companies with combustion engines such as diesel engines, which they wanted to stay with, even at the cost of cheating software. The intensification of the trend towards path dependency was by no means only determined by the interest in fast and secure corporate profits. Socio-political and trade union interests were also brought into play in order to secure millions of jobs with the old fossil fuels.

Once again, those responsible in politics and business preferred to stay on the comfortable and profit-promising tracks rather than switch to new innovations in good time. Every shareholder knows that a good portfolio needs to be diversified in order to make a good profit if individual shares fail. Future technologies are also bets on the future.

Joseph Schumpeter had already recognised the danger of market forces becoming entrenched in well-trodden paths and spoke of the 'creative destruction' of capitalism [5]. According to Schumpeter, innovations break up these incrustations and are the driving forces of the market economy that secure prosperity (Chapter 3).

In view of the climate and environmental crisis, sustainable innovations such as renewable energies are required, as well as nuclear energy with new types of nuclear power plants and fuel rods with reduced radiation. The warning signals of critical control parameters are unmistakable in all climate models and do not tolerate delay in the path dependency of convenience.

The central thesis and demand of this book is therefore the call for an innovation portfolio.

In this book, we speak of an innovation portfolio that can be used to manage the future. An innovation portfolio is made up of the current initial, bridging and future technologies, which are interlinked and interdependent and are in constant phase transitions. An innovation portfolio is therefore dynamic.

In the complex system dynamics of the economy and society, it is innovation and competition that trigger new markets, developments and changes. For Schumpeter, it is entrepreneurs who utilise opportunities with new innovations, raise capital and take risks. In biological evolutionary models, the mathematical system dynamics are explained by mutations and selections instead of innovations and markets. In view of the global environmental and climate crisis, all innovations must be measured against the standards of sustainability, such as those set out in the United Nations Sustainable Development Goals (SDGs) [4].

Information and energy technologies, which are directly linked, are key to sustainable innovation. On the one hand, new technologies for renewable energies are needed to get away from the environmental and climate impact of fossil fuels. Examples include solar and wind energy, water and hydrogen, and fusion and nuclear energy, which need to be combined in hybrid supply networks. In this context, we also speak of a dynamic innovation portfolio in which the diverse energy systems are to be combined or 'hybrid' as starting and bridging technologies. As with shares, the diversity of the portfolio protects against crashes in the event of one-sided dependencies. In this sense, future energy technologies are also bets on the future. However, dynamic innovation portfolios for energy technologies should not be geared unilaterally towards optimising profits but towards sustainability.

Efficient control of these processes is not possible without artificial intelligence (AI), information and communication technology. In addition, the gigantic growth of data and information in complex civilisations requires increased computer performance. This computing power must in turn be energy-efficient in order to promote sustainability. Therefore, as with the various energy technologies, different IT and computer technologies must also be considered — from classic computers with artificial intelligence and machine learning to quantum computing and neuromorphic computers modelled on the human brain.

Generative AI is currently playing a key role with innovative breakthroughs such as ChatGPT on the American side and DeepSeek on the Chinese side. Conclusions must be drawn from the advantages and disadvantages of these programmes.

Digital AI systems are themselves examples of complex dynamic systems that are realised in machine learning using neural networks. In terms of their architecture, they are reminiscent of the networking of neurons and synapses in natural brains. In reality, however, they are only simulated on digital computers according to the traditional scheme of a von Neumann architecture. This means that computing steps have to be processed one after the other (sequentially) and, for example, the control and memory units for data processing are separated. This makes it necessary to constantly 'shovel' data back and forth between processor and memory for all computing steps, which means both a large energy requirement and environmental pollution due to additional heat dissipation.

In contrast, these separations are eliminated in the neural networks of natural brains. Natural brains process information in parallel, with storage and information processing integrated in nerve cells (neurons). This results in unprecedented energy efficiency and at the same time unrivalled intelligence in natural brains. Neuromorphic computing with brain-like architectures is therefore highly attractive, albeit on the technical hardware of silicon and other nanoscale networks and (not yet) biological tissues. Here, too, mathematically complex system dynamics form the basis.

Finally, this book considers quantum computing as another future technology of information and communication technology. As in the example of the various energy technologies, it ultimately boils down to a hybrid information and communication technology in which the various computing paradigms are integrated in order to utilise their different advantages and disadvantages for a sustainable overall solution.

Innovations today are in a global competition between different value systems. Hegel and Marx each described this dynamic in their own way as 'dialectics'. Beyond all ideologies, we can now model these dynamics mathematically and in a much more differentiated way with phase transitions of complex dynamic systems, test them empirically, make limited forecasts and use them as early warning systems.

However, value systems are not measured only in economic terms. Behind the tanks, missiles and nuclear bombs of the superpowers, behind the Gross Domestic Product (GDP) figures and other economic indicators of their national economies, there are above all different ethical and legal value systems of the world powers that determine their actions and influence their innovation dynamics. During the Cold War until 1990, the centres of power were bi-polar. A metastable balance prevailed in the shadow of the atomic bomb between an Eastern and Western bloc.

After a brief interim phase in which one world power seemed to be spreading its dominance, we are now in a phase transition to a multi-polar world in which many old and new centres of power are emerging and competing with each other. As with the shifting continental plates, friction and violent eruptions occur at the edges.

> It is important to find a multi-polar dynamic equilibrium that can balance out local changes.

Are there generally binding rules according to which the competition between value systems should be played out? Since the World Wars of the 20th century, combined with the development of modern weapons technology up to apocalyptic self-destruction through nuclear weapons, wars can no longer be regarded as a 'continuation of politics by other means' (Clausewitz). In a densely populated and highly developed civil society, the effects of even a 'special military operation' [5] are so monstrous and ultimately uncontrollable that the use of military means should be prohibited from the outset. The 'frictions' with human and material 'costs' are so great today that in the end there can only be losers. Behind all the human dismay at the millions of individual sufferings and misery is ultimately the complexity of modern societies, which drives up the 'price' of war.

Complex system dynamics do not automatically lead to the development of a peaceful civilisation, as Hegel assumed in his belief in an evolving world reason. A study of the mathematics of dynamic systems shows how it is possible to fall into chaos and destruction. Moreover, the complex dynamics of a civilisation should not be confused with fluid dynamics in physics. Instead of molecules, conscious individuals are the elements of complex systems that pursue intentions and interests. Therefore, normative regulation through law and its enforcement is required.

However, mathematical analysis and modelling of complex system dynamics is not enough to overcome the conflicts and challenges of the 21st century. It also requires the justification and enforcement of norms that serve to orientate and set goals for human decision making and action. The overarching norm for the 21st century is the imperative of sustainability in order to secure the future of this planet (Chapter 6). Mathematical analyses of complex ecological, economic and social models provide the justification for this demand. They should help to avoid a relapse into the monstrosity of the 20th century with its terrible wars.

References

[1] Mainzer, K. (2007), *Thinking in Complexity. The Computational Dynamics of Matter, Mind, and Mankind*, 5th extended edn., Springer: New York.

[2] von Clausewitz, C. (2016), *On War*, Complete edition of the eight books. Edition Holzinger: Berlin.

[3] Arthur, W. B. (1989), Competing technologies, increasing returns, and lock-in by historical events, *The Economic Journal*, 99, pp. 116–131.

[4] Schumpeter, J. A. (2005), *Capitalism, Socialism and Democracy*, UTB: Stuttgart, 7th chapter: 'The opening of new, foreign or domestic markets and the organisational development from the craft workshop and factory to such corporations as U.S. Steel illustrate the same process of industrial mutation — if I may use this biological term — that is constantly revolutionising the economic structure from within, constantly destroying the old structure and constantly creating a new one. This process of 'creative destruction' is the essential fact of capitalism. This is what capitalism consists of and this is what every capitalist structure must live in'. [5] In order to create globally sustainable structures, the member states of the United Nations have set themselves 17 goals by 2030, which are set out in the 2030 Agenda for Sustainable Development.

[5] Official name of the war of the Russian Federation for the war against Ukraine.

Chapter 2

Dynamics of Complex Systems

Complexity research brings together different approaches drawn from various sciences [1]. On the one hand, the sciences today are highly differentiated and specialised in a complex variety of individual disciplines. On the other hand, the sciences themselves have to deal with highly complex systems in nature and society — from complex atomic, molecular and cellular systems in nature to complex social and economic systems in society. Complexity research deals with the interdisciplinary question of how the interaction of many elements of a complex system (e.g. molecules in materials, cells in organisms or people in markets and organisations) can lead to order and structure as well as chaos and collapse. Complexity research aims to recognise chaos, tensions and conflicts in complex systems and to understand their causes in order to gain insights into new potentials for shaping the systems.

To this end, new basic concepts, measurement methods, models and algorithms are introduced. In this way, complex orders can be characterised by order parameters. Like chaos and decay, order arises in critical states that depend sensitively on the control parameters of a system or are self-organising. These characterised states are often called attractors, as the dynamics of a system are drawn into a vortex, as it were. Complex patterns of time series and other criteria are used to recognise critical situations from measurement data in advance and to take precautions in good time. Computer models play a decisive role here. The dynamics of complex systems in nature and society have only been able to be analysed in simulation models for a few years now thanks to the increased computing capacity of computers.

Definition of dynamic systems

It is not only the mass of the data but also their diverse interactions that make calculations difficult. Are we dealing with the fundamental limits of computability?

In general, a dynamic system consists of a set of elements that change over time [2]. The interactions between the microscopic states of the elements determine the overall macroscopic state of the system. In a planetary system, for example, the state of a planet at a point in time is determined by its position and speed. However, it can also be the state of motion of a molecule in a gas, the state of excitation of a nerve cell in a neuronal network or the state of a population in an ecological system. The dynamics of the system, i.e. the change in the system states over time, is described by time-dependent equations (e.g. differential equations). In the case of deterministic systems, each future state is uniquely determined by the present state.

Linear dynamics

A simple example is a harmonic oscillator. With a mass attached to a spring, a small deflection leads to a small oscillation, while a large deflection causes a large oscillation as an effect. In linear systems, causes and effects are proportional (Fig. 1(a)). Mathematically, we then obtain an equation of the form $f(x) = c \cdot x$ with x values (e.g. for the deflection of a body on a spring), the dependent function values $f(x)$ (e.g. the force on the body) and a proportionality constant c (e.g. depending on the material of the spring). As this equation represents a straight line with the gradient c in the coordinate system, it is called linear.

State space of dynamic systems

A solution to these equations of motion can be represented graphically as a time series of location as a function of time (Fig. 1(b)). A regular oscillation along the time axis (e.g. in a pendulum)

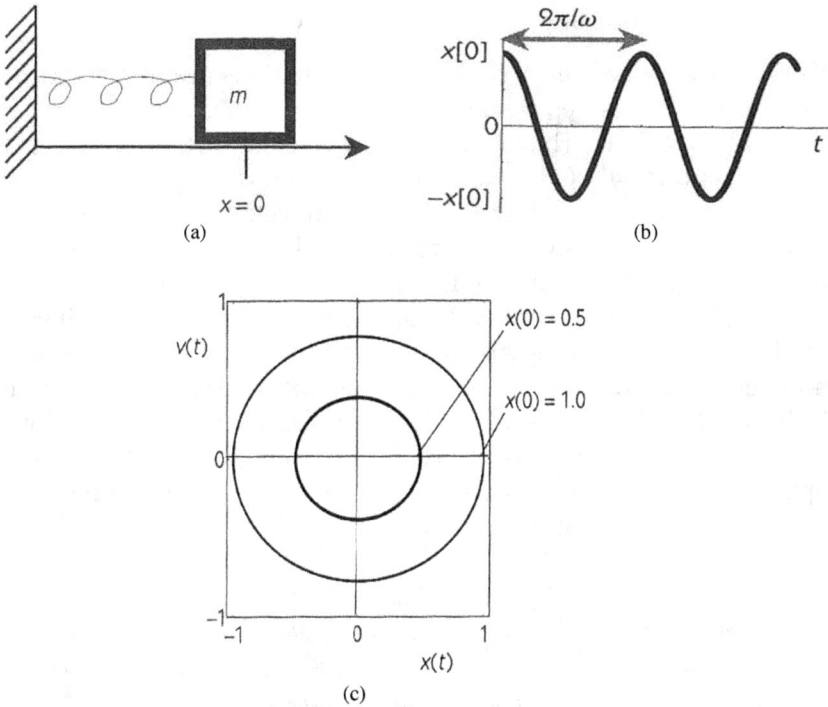

Fig. 1. Dynamics of an oscillator (mass of a spring) in an experimental space (a) as time series of measuring dates (b) and in a state space (c).

corresponds to a closed path (trajectory) in the state space, in which all states of motion of the dynamic system are represented as points. The points, i.e. the states of motion, are represented by the two coordinates of the location and the velocity of the mass at a point in time (Fig. 1(c)).

Due to the spring oscillation, the mass always returns to its initial state in this case and therefore forms a closed curve in the state space. Different initial states lead to different state curves [3]. In the state space, we therefore fully recognise the dynamics of a linear oscillator for all possible situations. In this case, a causality analysis is not only completely feasible but also calculable. This is the world that Laplace believed in: under these conditions, all of nature would be computable.

Non-linear dynamics

We know from mathematics that linear equations are easy to solve. However, non-linear equations, which represent geometric curves, do not always allow arbitrarily accurate calculations, even with our best computers [4]. One example is the multi-body problems of celestial mechanics, in which more than two celestial bodies interact through gravity. They generate feedbacks that correspond to non-linear equations of motion of the planets. The French mathematician, theoretical physicist and philosopher Henri Poincaré (1854–1912) was the first to show that chaotically unstable orbits can occur in a non-linear multi-body problem that depends sensitively on their initial values and cannot be predicted in the long term. In 1908, alluding to Laplace, he wrote in one of his philosophical books, aptly contradicting the belief at the time in a totally determined world:

> *If we knew the laws of nature and the initial state exactly, we could predict the state of the universe at any further point in time. But even if the laws of nature no longer held any secrets for us, we could only approximately determine the initial conditions.*
>
> *If this allows us to specify the following states with the same approximation, we say that the behaviour has been predicted, that it follows laws. But this is not always the case: it can happen that small differences in the initial conditions result in large differences in the outcome (...) prediction becomes impossible and we have a random phenomenon* [5].

Multi-body problems and the limits of computability

At school, we learn about the interaction of celestial bodies as a two-body problem, using the example of a planet that moves around the Earth in an elliptical orbit according to Kepler. The solutions to the corresponding equation of motion can be calculated according to Newton. Now, the question arose for the interaction of three celestial bodies such as the Sun, Moon and Earth or even all celestial bodies as a general multi-body problem. This question was closely linked to the stability of the solar system.

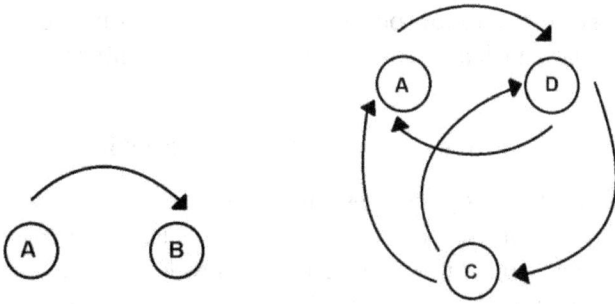

Fig. 2. Two- and many-body problems.

Poincaré showed that the equations describing the interactions of three bodies generally have no solutions that can be calculated using algebraic formulae and integrals (i.e. using standard solution methods) (Fig. 2). However, in 1912, the mathematician Karl Sundmann was able to specify an infinite convergent series for three bodies, the terms of which could at least in principle be totalled for a solution. However, this convergence is extremely slow. In 1991, the Chinese mathematics student Qiu-Dong Wang was able to generalise the solution with power series for an arbitrary multi-body problem [6]. However, for a practical solution, millions of terms of the power series would have to be analysed. This method is therefore out of the question for practical calculations. One could argue that an infinite Laplacean spirit could calculate these processes. In any case, the exact calculation for terrestrial supercomputers is (so far) practically impossible.

KAM theorem and the limits of computability

In the spirit of Poincaré, Andrei N. Kolmogorov (1954) [7], Vladimir I. Arnold (1963) [8] and Jürgen K. Moser (1962) [9] finally proved their famous KAM theorem: Temporal trajectories in the state space of classical mechanics are neither completely regular nor completely irregular, but depend sensitively on the chosen initial conditions.

Tiny deviations from the initial data lead to completely different developmental trajectory. Therefore, the future developments in a chaotic system cannot be computed in the long run, although they are mathematically well defined and determined.

Recursive procedures and difference equations*

Instead of continuous processes, discrete processes can also be analysed as change of system states in temporal steps through difference equations [10]. An example is the developmental dynamics of a population. The quantity x_n of the population in the nth year determines the quantity of the succeeding population x_{n+1} in dependence of a reproduction rate r. Mathematically, that is expressed through a linear growth function $f(x) = r \cdot x$. It corresponds to a recursive function of succeeding generations $x_{n+1} = f(x_n) = r \cdot x_n$, which refers to the previously calculated value x_n for the calculation of a function value $f(x_n)$. Intuitively, population values are generated such as nested Russian dolls. The computational complexity depends on the recursive steps and the reproduction rate r. Starting with initial population x_0 leads to exponential growth x_0, $r \cdot x_0$, $r^2 \cdot x_0$, $r^3 \cdot x_0$, ... because of x_0, $f(x_0)$, $f(f(x_0))$, $f(f(f(x_0)))$ ... .

If resources (e.g. nutrition of a population) are restricted, a negative feedback of growth must be considered. It is stronger the larger the generation in question is. Let the largest population value be standardised with 1. In this case, the feedback is expressed by $1 - x$. Therefore, the mathematician and sociologist Pierre Francois Verhulst (1804–1849) suggested the quadratic (non-linear) growth function $f(x) = r \cdot (1 - x)x = r \cdot (x - x^2)$. The corresponding recursive function $x_{n+1} = f(x_n) = r \cdot (1 - x_n)x_n$ of succeeding generations creates a surprising variety of growth patterns depending on the reproduction rate r. The reproduction rate is the control parameter of this dynamical system.

Time series and complexity degrees

The complexity degrees of the Verhulst dynamics are represented in the time series of succeeding generations (Fig. 3). Weak growth r begins at first with an exponential curve, which then leads to a plateau of equilibrium. The plateau of this so-called logistic curve corresponds to an equilibrium point. This means that from a quantity x^*, the population does not change any longer and remains

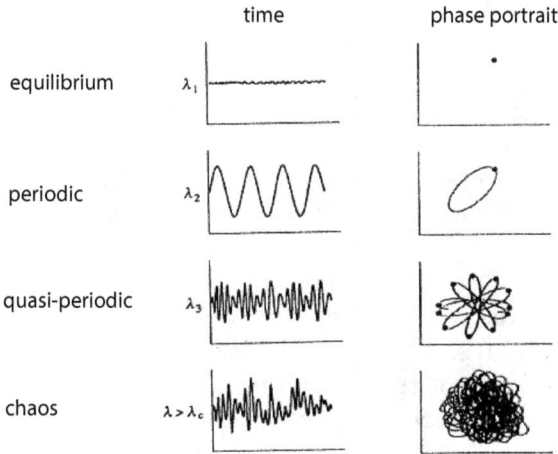

Fig. 3. Complexity degrees of time series and attractors.

fixed at point $f(x^*) = x^*$, for all succeeding generations. In nature, a population adapts to the conditions of the environment. For stronger growth, an oscillation between two population quantities is generated. A practical example is overpopulation with insufficient nutrition leading to oscillation. In the case of much stronger oscillation, several population quantities emerge, between which the population curve oscillates. In this way, quasi-periodic developmental patterns are generated, which are rhythmically repeated. In the case of very strong growth, completely irregular (non-periodic) chaotic oscillations emerge at a critical value of the growth parameter, which sensitively depend on tiny changes of the initial data (Fig. 3).

Attractors in the state space

In the state space, you can see how the state development (trajectory) of a dynamical system aims at characteristic patterns (attractors): an attractor is a state, in which a dynamical system is attracted for the long term [11]. An equilibrium state corresponds to a fixed-point attractor, which no longer changes in time ('remains fixed'). In the state space, all developmental lines (trajectories) converge to this point as a final state. Linear systems only have fixed-point attractors. Non-linear systems also have limit cycles in which states are periodically repeated or in the case of turbulence chaos attractors in which the developmental lines completely irregularly and non-periodically condense in a limited region of the state space (Fig. 3). In the case of random developments, all correlations decay into independent events and fluctuate irregularly over the entire state space. Dynamical complexity and chaos are situated between complete regularity (as in the case of linear systems) and random.

In the first case, the Verhulst dynamics aims at a fixed-point attractor. In the second case, the Verhulst dynamics fluctuates between two states. In the third case, even densely neighboured initial values lead to irregularly diverging trajectories after a few steps of iteration. In the computer model, tiny changes of digitalised initial dates lead to exponentially growing computational time of future dates, which practically excludes long-term predictions.

Random and chaos

Nevertheless, it holds that in deterministic chaos the dynamics is completely determined through a non-linear growth law. But long-term effects are practically not predictable, as the computational amount grows exponentially because of the sensible dependence of the initial data. Contrary to chaos, random developments are in principle (also short term) not predictable, because (such as in the case of a fair coin toss) all events are independent. In everyday life, random is only a mathematical fiction because there is no perfect fair coin as long as we are in classical dynamics. In the quantum world, random 'quantum jumps' belong to the natural laws.

Fluid dynamics and stochastic equations

In dynamic systems, degrees of complexity can be distinguished if, as in the case of fluid dynamics, we cannot know the individual micro states due to the enormous number of system elements (e.g. gas molecules) [12]. The macro state of the system is then determined by a probability distribution of micro states of the system elements. The dynamics, i.e. the temporal development of this distribution function, is described by a stochastic equation. Let us consider the flow patterns in a river behind an obstacle as a function of the increasing flow velocity. Initially, the flow has a homogeneous flow pattern behind the obstacle. It corresponds to a homogeneous state of equilibrium as a fixed-point attractor. As the flow velocity increases, individual vortices are formed. They correspond to periodic cycles. If the flow velocity is increased further, the vortices combine to form vortex patterns that correspond to quasi-periodic cycles. Finally, the flow pattern changes to non-periodic and irregular vortex patterns, which correspond to a chaos attractor (Fig. 3).

Another example is provided by non-linear flow dynamics in meteorology, according to which the slightest local changes, such as a small unnoticed vortex on the weather map, a spinning leaf or the flap of a butterfly's wings, can trigger global chaotic changes in the general weather situation. In chaos theory, this is referred to as the 'butterfly effect' after the meteorologist and mathematician Edward N. Lorenz. Despite the high computing capacity of today's computers, only short-term forecasts are possible. In order to measure the complexity of a time series and thus of a non-linear dynamics, we can, for example, determine the degree of non-periodicity or the sensitive dependence of a dynamics on its initial data. For example, the so-called Lyapunov exponents can be used to measure whether and how the trajectories drift apart in the state space in order to determine the degree of sensitive dependence (butterfly effect).

Micro and macro level of complex systems

The basic idea of complexity research is always the same across disciplines: only the complex interactions of many elements create new properties of the overall system that cannot be traced back to individual elements. For example, a single water molecule is not 'moist', but a liquid is due to the interactions of many such elements. Individual molecules are not 'alive', but a cell is due to their interactions. In systems biology, the complex chemical reactions of many individual molecules enable the metabolic functions and regulatory tasks of entire protein systems and cells in the human body. In complex dynamic systems, we therefore distinguish between the micro level of the individual elements and the macro level of their system properties. This emergence of new system properties is also known as 'self-organisation'.

Control and order parameter

Synergetics emerged in the 1970s in the statistical physics of non-equilibrium systems, initially investigating physical systems such as lasers. This example of self-organisation far from thermodynamic equilibrium was used to develop the essential principles of synergetics, such as the order parameter, the enslavement principle and the connection with the theory of phase transitions. The collective self-organisation of the atoms in the laser is a synergetic effect that is brought about by a critical phase transition of the entire system. Control parameters indicate the critical values at which the phase transitions take place. Mathematically, a phase transition means a considerable simplification: instead of more than 10^{80} equations for all atoms and photons at the micro level, a few equations are sufficient for order parameters that determine ('enslave') the overall macroscopic behaviour of all elements. It is therefore important to identify the order parameters.

A linear stability analysis is used for this purpose, in which the (analytically) unsolvable non-linear equations are reduced to a linear equation [13]. The non-linear terms are neglected in a Taylor expansion of the non-linear equation in order to obtain a solvable linear equation. In the solutions of this linear equation, terms (eigenvalues) can be distinguished that correspond to the stable or unstable behaviour of modes. In the case of unstable behaviour, the oscillations of one mode build up to such an extent that they are transmitted to the other modes. The corresponding terms are therefore labelled as order parameters because this mode 'imposes' its order on the other participants. These calculations are also referred to as an adiabatic approximation. Approximative computability is therefore achieved by linearising a complex problem.

This means that the uniform fundamental mode, which is formed by phase transitions in the laser light system, can be calculated using the dominant order parameter alone. This simplifies the differential equations so that they can be solved. From a physical point of view, the atoms in a laser instantly follow the specifications of the dominant order parameter according to the method of adiabatic approximation in the sense of synergetics.

Pattern and structure formation

The contribution of the British logician and computer pioneer Alan Turing (1912–1954) was fundamental for computability issues with computers. Shortly before his death, Turing had also worked on structure and pattern formation in nature [14]. What was remarkable about his work was that he assumed only two stable subsystems (e.g. cells with a few molecules), which became unstable after dissipative coupling. As his equations of dissipative interaction were linear, no complex structure and pattern formation of the overall system could be expected. It was not until 1974 that the American mathematician (and winner of the Fields Medal) Steven Smale (UC Berkeley) generalised Turing's approach for non-linear interactions and proved that in this case even stable cells with dissipative coupling lead to oscillating solutions, i.e. complex structure and pattern formation.

Local activity

This raises the question of what must be the common prerequisite for both unstable and stable cells so that their (non-linear) interaction leads to the formation of structures and patterns. Based on Turing and Smale's special case, a spatial system consisting of any number of identical elements ('cells') that can interact with each other in different ways (e.g. physically, chemically or biologically) was presented in [15]. Such a system is called complex if it can generate non-homogeneous ('complex') patterns and structures from homogeneous initial conditions. This pattern and structure formation is triggered by the local activity of its elements. This applies not only to stem cells during the growth of an embryo but also to transistors in electronic networks, for example. We call a transistor locally active if it can amplify a small signal input from the energy source of a battery into a larger signal output in order to generate non-homogeneous ('complex') voltage patterns in switching networks.

No radios, televisions or computers would be able to function without the local activity of such units. Important researchers such as the Nobel Prize winners Ilya Prigogine (chemistry) and Erwin Schrödinger (physics) were still of the opinion that a non-linear system and an energy source were sufficient for structure and pattern formation.

However, the example of transistors already shows that batteries and non-linear switching elements alone cannot generate complex patterns if the elements are not locally active in the sense of the amplifier function described.

The principle of local activity is of fundamental importance for pattern formation in complex systems. In [16], non-linear differential equations that describe reaction–diffusion processes between the cells are analysed.

In general, a cell is called locally active if there is a small local input at a cellular equilibrium point that can be amplified to a large output with an external energy source. A cell is called locally passive if there is no equilibrium point with local activity. It can be shown that systems without locally active elements cannot, in principle, generate complex structures and patterns.

Edge of chaos

The existence of an input that triggers local activity can be systematically tested mathematically using certain test criteria [17]. These criteria initially describe the locally active and unstable cases, which are also covered in the linear stability analysis of synergetics. (There, the unstable modes are labelled as order parameters.) However, the criteria also take into account the case of Turing and Smale, in which locally active and (asymptotically) stable cells lead to structure and pattern formation after dissipative coupling.

Stable cells are to a certain extent mathematically 'dead' and are only brought 'to life' by dissipative coupling if they have the potential for local activity. In this case, we therefore also speak of the 'edge of chaos', where, for example, 'dead' chemical substances can trigger life processes after dissipative interaction. Order parameters therefore not only correspond to local activity in unstable states but also in stable states, which ultimately determine the entire system. This is a central extension and correction of the original approach of structure and pattern formation in complex dynamic systems.

References

[1] Mainzer, K. (Ed.) (2009), Complexity, *European Review* (Academia Europaea), 17(2), S. 219–452. Cambridge University Press: Cambridge.

[2] Mainzer, K. (2004), Dynamical systems, in Scott, A. (Ed.), *Encyclopedia of Nonlinear Science*, Fitzroy Dearborn: London, S. 240–241.

[3] Kaplan, D. and Glass, L. (1995), *Understanding Nonlinear Dynamics*, Springer: New York, Kap. 5.

[4] Mainzer, K. (2005), *Symmetry and Complexity. The Spirit and Beauty of Nonlinear Science*. World Scientific: Singapore.

[5] Poincaré, H. (1908), *La Science et Méthode*, Paris, E. Flammarion.

[6] Wang, Q. (1991), The global solution of the n-body problem, *Celestial Mechanics and Dynamical Astronomy*, 50, S. 73–88.

[7] Kolmogorov, A. N. (1954), On the conservation of conditionally-periodic motions for a small change in Hamilton's function, *Doklady Akademii Nauk USSR*, 98, S. 525.

[8] Arnold, V. I. (1963), Small denominators II. Proof of a theorem of A. N. Kolmogorov on the preservation of conditionally-periodic motions under small perturbation of the Hamiltonian, *Russian Mathematical Surveys*, 18, S. 5.

[9] Moser, J. (1967), Convergent series expansions of quasi-periodic motions, *Mathematische Annalen*, 169, p. 163.

[10] Mainzer, K. (2008), *Komplexität*, UTB-Profile: Paderborn.

[11] Mainzer, K. (2007), *Thinking in Complexity*, 5th extended edn., Springer: Springer.

[12] Nicolis, G. and Prigogine, I. (1987), *Die Erforschung des Komplexen*, Piper: Munich.

[13] Haken, H. (1990), *Synergetics. An Introduction*, 3rd edn., Springer: Berlin.

[14 Turing, A. M. (1952), The chemical basis of morphogenesis, *Philosophical Transactions of the Royal Society London, Series B*, 237, S. 37–72.

[15] Mainzer, K. and Chua, L. (2013), *Local Activity Principle*, Imperial College Press: London.

[16] Mainzer, K. and Chua, L. (2013), *Local Activity Principle*, Imperial College Press: London, Chapter 1.1.

[17] Mainzer, K. and Chua, L. (2013), *Local Activity Principle*. Imperial College Press: London, pp. 12–15.

Chapter 3

Innovation, Evolution, Disruption

Computability of sociodynamics

Risks, uncertainty and probability are typical of complex societies in which humans live [1]. In order to recognise trends and attractors of future developments in such complex systems, a distinction is generally made between the micro level of the system elements and the macrodynamics of the overall system. This is where the modelling concept of sociodynamics comes in Refs. [2, 3]. Recently, the misleading terms 'sociophysics' and 'econophysics' (from economics and physics) have also been used. They suggest a social–scientific physicalism, according to which social processes are traced back to the laws of physics, as in the 'sociophysics' of the 19th century [4, 5]. In fact, this is the mathematical formalism of complex dynamic systems, which is independent of physical variables and constants and is interpreted using suitable social and economic state variables [6].

Methodologically, the micro level of individual decisions of single people is distinguished from the macro level of collective processes. There is only a formal analogy with thermodynamics, as probabilistic collective developments are also modelled by stochastic differential equations (e.g. master equations as in thermodynamics) [7]. However, the underlying social macro states (socioconfigurations) are measured using social science methods. Each component of a socioconfiguration refers to a subpopulation with a characteristic behaviour vector.

In a socioconfiguration, a distinction is made between material variables (e.g. prices on the stock exchange, property values of a company, production volume) and personal variables (e.g. job details, biographical data, attitudes, moods), which are assigned to different subpopulations. In this way, different departments can be distinguished in a company, which are determined by the status of different buildings, company tasks, groups of people and similar. The socioconfiguration describes the overall state of the company at a point in time. Stochastic assumptions about probable changes in a time interval (phase transitions) can be made for its individual segments [8]. For example, how likely is it that an employee will move from one department to another? How likely is it that his or her attitude towards a certain topic will change?

These assumptions are made on the basis of static surveys. If we visualise the individual partial states of the sections in a flowchart, partial transitions are constantly taking place, which lead to phase transitions of all the socioconfigurations at the macro level. The temporal change of the socioconfigurations describes the dynamics of the entire social system (e.g. a company) and can be modelled by a stochastic differential equation (e.g. master equation).

The situation is known from statistical physics [9]: In a Brownian motion, a particle is pushed forwards or backwards at random. Continuous hopping processes change the probability of finding a particle at a certain place at a certain time. Its movement can be described by an equation that determines the temporal change of a probability distribution. In statistical physics, the so-called master equation is used to calculate the temporal change in a probability distribution during a random process. The master equation is of interdisciplinary importance for all random processes in physics, chemistry, biology, economics and social sciences [10]. It can be applied to probability distributions of molecules, cells, organisms, economic agents or citizens of a society. It centres on the question of how a probability distribution changes over time under certain conditions.

In computer graphics, for example, the changing migration flows of two populations can be depicted as different attractors analogous to river dynamics — from ghetto formations (point attractors) to oscillating and chaotic states. However, social and economic factors and their interactions are taken into account and not energetic interactions.

At the micro level, we cannot predict individual decisions. However, we do observe the typical local activity of individuals, which we have already observed in biology with cells in organisms and neurons in brains. Local activity means amplifying input signals and translating them into actions that, together with other individuals, form new clusters at the macro level (e.g. markets, companies, institutions). Behaviour at the 'edge of chaos' is of particular interest, as complex structures can be expected here. One example is a start-up in which originally uninvolved stakeholders of a university come together to jointly develop an innovation [11].

If one knew the stochastic equations of these individuals, one could determine their parameter spaces and, in principle, calculate these cluster formations. In fact, possible scenarios of collective trend developments can be recognised and simulated at the macro level under certain secondary conditions (control values), even if the equations of motion of the individuals are not known at the micro level.

Sociodynamics is not based on a new type of causality. Social systems are also non-linear dynamics of complex systems that take into account a variety of feedback from the simultaneous interactions of many elements. However, we generally have no equations of motion for the individual behaviour of the system elements at the micro level. People are not molecules or cells. Nevertheless, their political preferences, for example, generate collective voting trends that have an effect on the voting behaviour of individuals in a similar way to flow patterns.

One could consider using equations of individual brain dynamics as a basis at the micro level. However, this approach would not be practicable due to the complexity of such equations. Social scientists therefore assume random fluctuations at the micro level and work at the macro level with statistical distribution functions whose dynamics are modelled with stochastic equations [12].

Internet of things as a complex dynamical system

Today, the Internet is a dominant, complex system of network nodes that can interconnect to form patterns. In the Internet of Things, devices of all kinds (e.g. smartphones, cars, houses) are equipped with sensors to communicate with each other. The result is an unmanageably large amount of 'big data' [13]. As in road traffic, critical control parameters (e.g. data density, transmission capacity) can lead to data congestion and data chaos. Mathematically, these networks are complex systems with non-linear dynamics, as we have already seen in cells, organisms and brains. The non-linear side effects of these complex systems can often no longer be controlled globally. Local causes can build up to unforeseen global effects due to non-linear interactions.

We therefore also speak of systemic risks, which have no individually identifiable causes but are made possible by the system dynamics as a whole. Even in seemingly (asymptotically) stable states, a crash is possible at any time on the 'edge of chaos'.

Terrorist threats are an example of this in social systems.

Our technology is becoming more autonomous in order to solve the tasks of an increasingly complex civilisation. Individual people can no longer understand the organisational systems required for this. The downside of the increasing autonomy of technology is that it is becoming more difficult to control: machines and devices have always been developed in the engineering sciences with the intention of being able to control them. But how can systemic risks of complex systems be avoided?

A look at evolution shows that autonomous self-organisation and control have complemented each other there. In diseases such as cancer, however, this balance is disturbed: a cancerous tumour is a self-organising organism that develops its own interests and fights for its survival, so to speak, but does not realise that its own host organism will perish as a result. Complex systems therefore need control mechanisms in order to find a balance — in organisms, financial markets and politics. It is necessary to know the principles and limits of the algorithms used [14]. These mega systems of micro and macro worlds develop their own non-linear dynamics. However, they should remain service systems for the benefit of man and his civilisation.

Complexity management and computability

Complexity and crisis management are successful when we understand the non-linear dynamics of complex systems [15]. It is therefore important for a company to find out how close it should move to instabilities with their typical random fluctuations in order to trigger bursts of innovation and avoid slipping into disintegration, disorientation and chaos. At the 'edge of chaos', on the other hand, 'local activity' can lead to extraordinary creativity. 'Disruption' is therefore not a mystical intrusion into world history, but rather a mathematical possibility that opens up complex system dynamics. In the theory of complex dynamic systems, global trends can be modelled by a few statistical distribution variables. For example, we do not need to know the actual micro behaviour of each individual driver in order to be able to predict macro behaviour such as stop-and-go waves or gridlock for certain traffic densities. Intelligent traffic management systems must learn to recognise such trends in good time from statistical density patterns and adapt to the traffic flow. Intelligent management must also learn to deal sensitively with instabilities and random fluctuations and set suitable framework conditions so that the desired business dynamics can organise themselves.

But companies, it will be argued, are systems of people with feelings and consciousness, not mindless atoms or molecules. However, global opinion trends also arise in social groups, on the one hand, through collective interaction between their members (e.g. communication). On the other hand, global trends have a feedback effect on the group members, influencing their micro behaviour and thereby reinforcing or slowing down the global system dynamics. Such feedback loops between the micro and macro dynamics of a system enable learning effects in the company, such as anti-cyclical behaviour, in order to consciously counteract harmful trends.

Local activity and innovation dynamics

Local instability in complex social systems not only leads to extreme events with chaos and catastrophic collapse of the entire system, but new social, economic and technological structures also emerge, triggered by new inventions and innovations.

> Inventions are just new ideas and realisations of technical concepts. Innovations are inventions that become established on the markets or are accepted by consumers.

Are they caused by external influences or by spontaneous internal fluctuations in system dynamics? Natural and social systems are often subject to external disturbances with different amplitudes. It is sometimes difficult to decide whether an observed extreme and exceptional event is due to a strong exogenous shock or to the internal (endogenous) dynamics of the complex system.

In the past, most people believed that inventions and innovations were spontaneous events of ingenious minds that were not subject to any causal dynamics, like mysterious miracles. In this case, inventions and innovations are random events without any chance of human influence.

In fact, however, a careful analysis of innovation dynamics shows that the emergence of new ideas, inventions and innovations is due to internal laws of non-linear dynamics and the principle of local activity. If inventions and innovations are endogenous events, then their emergence can be influenced by analysing the dynamics of innovation and changing its boundary conditions. In short, in this case we can learn and improve the chances of innovation.

In general, local activity in a dynamic system means that a local unit of the system amplifies a low energy input from outside during a certain period of time and transforms it into a sum of high energy, forming complex structures and patterns. Applied to innovation dynamics, investments in creative innovation centres, e.g. in research centres, universities or companies, are the inputs that are transformed into new inventions or innovations and change the economy and society. Innovation dynamics is therefore an excellent example of the principle of local activity.

Schumpeter's theory of innovation dynamics

A forerunner of modern innovation dynamics was the Austrian economist Joseph F. Schumpeter, who recognised the importance of technological discontinuities in economic history [16]. Schumpeter argued that 'evolution is lopsided, discontinuous, disharmonious by nature ... studded with violent outbursts and catastrophes ... more like a series of explosions than a gentle, though incessant, transformation'. According to Schumpeter, the dynamics of innovation correlate with economic cycles. New ideas are constantly emerging. When enough ideas have accumulated, a bundle of innovations is realised through entrepreneurship. The accumulation process refers to the sum (mathematically the integral) of potential forces in the sense of the principle of local activity.

The innovation bundle initially develops slowly and then accelerates as the methods are improved. A logistic curve characterises the typical course of an innovation. The introduction of an innovation must be preceded by an economic investment. Investments stimulate demand. Increasing demand facilitates the spread of the innovation. Then, when all innovations have been fully utilised, the process slows down towards zero.

Schumpeter called this phenomenon the 'swarming' of innovations. In his three-cycle model, the first short cycle relates to the stocks cycle and innovations play no role. The following longer cycle is related to innovations. Schumpeter recognised the significance of historical statistics and related the evidence of long waves to the fact that the most important innovations like steam, steel, railways, steamships and electricity required 30–100 years to become completely integrated into the economy.

In general, Schumpeter described economic evolution as a technical progress in the form of 'swarms' which were explained in a logistic framework. A technological swarm is assumed to shift the equilibrium to a new fixed-point attractor in a cyclical way. The resulting new equilibrium is characterised by a higher real wage and higher consumption and output. Thus, Schumpeter's innovation dynamics can easily be interpreted in terms of sociodynamics with attractors. Innovation swarms at economical points of instability can be considered a global regime dominating long-term business cycles.

During the last decades, innovation dynamics has been studied as an endogenous part of economy. The message of these studies is that economic performance of companies not only depends on how business corporations perform but also on how they interact with each other and with the scientific and public sector. Innovation, knowledge creation and diffusion are considered interactive and cumulative processes with emerging innovatory patterns.

A system of innovation consists of a set of actors or subsystems such as firms, organisations and institutions interacting with use and diffusion of new knowledge. There are regional, national, European or global innovation systems with complex information networks on different scales. In any case, innovation systems are knowledge intensive [17].

Innovation systems not only use scientific knowledge but also technological, organisational and public knowledge. Knowledge means information and also experience, skill and wisdom. There is explicit knowledge which can be stored in data bases, and implicit knowledge expressed in personal know-how and skills.

According to Schumpeter, innovation dynamics is a historical and evolutionary process which cannot be understood in the neo-classical terms of economics. For example, path-dependence means that the development of an innovation follows a time-depending trajectory with a certain trend which cannot easily be left. If a company is engaged in, for example, nuclear energy, it is very difficult to change to windmill production, although it may be more acceptable under changing societal conditions. Path-dependence and multi-stability can be modelled in the framework of non-equilibrium dynamics.

From Schumpeter's 'creative destruction' to disruptive innovation

Following Schumpeter, 'disruptive' innovations are often described as 'destructive' because they replace old business models or technologies. The literal meaning is an abrupt or 'sudden' event. In mathematical system dynamics, the criteria of a phase transition must therefore be fulfilled. A popular example was the invention of the smartphone, which 'abruptly' replaced MP3 players, digital cameras, telephones and newspapers for many users.

In economics, the term 'disruption' goes back to Clayton M. Christensen [18]. Christensen explains 'disruptive innovations' using examples of innovations in new markets. It has since become a colourful buzzword in politics and the media, and is also often used in the social sciences in an inflationary and superficial manner. In fact, the phase transition of a technology is very complex. From the point of view of the scientific foundations of innovation, it is often a rather continuous and long-term process. This is clearly illustrated by the example of quantum technologies. Popularly, the transition from classical physics to quantum physics was presented as a 'break' with the classical worldview of physics. However, Einstein already spoke of standing 'on the shoulders of giants' and providing a 'small' but decisive impetus, which over the decades, together with many contributions from other authors, led to new theories and step-by-step developments.

So, there was an 'underground current' of basic research and clear indications that experts and specialists were well aware of it. It was only on the 'surface' of the media and popular information dissemination that the 'new' seemed surprising because it was not understood. With this background knowledge, quantum technologies from lasers to the current first models of quantum computers and quantum communication are no longer surprising and 'disruptive' for experts.

The term 'disruption' is only original for the development of business models for innovations. It is significant that customers are initially reserved about new technologies. Therefore, new customers must first be acquired who, for whatever reason, find the technology attractive. Basically, however, this also applies to the changing attitude of scientists in the natural sciences and has led to extensive discussions in the theory of science, e.g. Pierre Duhem, Thomas Kuhn, Imre Lakatos and others, on how 'disruptive' or continuous the theory transitions in scientific disciplines were and are Ref. [19].

One criterion for the transition of business models can be different demands on the quality of the new products.

As a result, digital photography was initially received with reservations by professional photographers, as its claim to high image quality was not initially fulfilled. It was sufficient for mass distribution on the Internet. However, the technical quality of digital photography quickly improved and led to the widespread replacement of analogue photography and the demise of entire industries associated with it (e.g. film production). Lovers of the old technology remained, but investors have to struggle with the risk of knowing when the critical 'control values' have been reached in order to make the 'path jump' in Schumpeter's sense in good time. In science, such a timely jump is associated with professional careers: If you jump too early, you end up in professional nirvana; if you jump too late, life punishes you.

The discussion about 'disruptive' technologies has led to a lot of confusion, especially in the high-tech sector. The adjective 'disruptive' is hype, so that anything and everything is labelled as such for advertising purposes. Because there are no clear criteria, there is also uncertainty as to whether a product is actually 'disruptive' and whether products labelled as 'disruptive' are successful. A study of start-ups shows how difficult it is to implement new technical ideas as innovations with a successful business model. This requires the bundling of expertise from science, technology, finance and corporate management. At a certain stage, psychology and media must also be taken into account, as the emotional and human component of users and customers should not be underestimated.

Agent-based dynamics of innovation network

In the following example of Günter Haag and Philip Liedl, the master equation approach of sociodynamics is applied to innovation dynamics with the complex interaction of labour force and capital formation, knowledge production and diffusion [20]. A master equation describes the evolution of the probability function, representing the transition probabilities for well-defined states of a dynamic micro-based system of actors.

Master equations model the time-depending change of macroscopic system structures which are linked to the micro level of interacting actors. Adaption processes and learning effects are synergetic effects which can be taken into account in this formal approach. Systems states are represented as socioconfigurations, including the individual transition probabilities of actors based on joint interaction effects. The master equation approach allows scenario-based computer simulations. The case studies illustrate that the innovation power of interlinked firms depend on spill-over effects between firms and the impact of the scientific system.

Agent-based complex networks of firms, universities and technology-transfer centres generate patterns of innovation activities. There are internal and external impulses for firms to innovate. Internal impulses come from the firm's own research, development and management decisions. External impulses arise from their products, trade fairs and contacts with universities and research centres. Spill-over effects describe the impact of different transfer

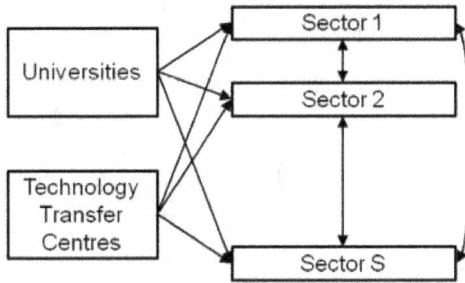

Socio-economic environment

Fig. 1.   Agent-based innovation network.

activities on the production of a specific firm. In Fig. 1 [21], a complex network of firms of different sectors interacting with universities and technology-transfer centres is illustrated. Positive spill-over effects happen if, for example, an innovation in one sector has a positive impact on other sectors. A negative impact on the development of products in other sectors means a negative spill-over effect.

Innovation process of firms*

Each firm $i(i = 1, 2, ...)$ is characterised by a production function $Q_i = f(K_i, L_i, I_i)$ with capital stock K_i, labour force L_i and the impact of innovations I_i. Additional to classical economics, the modified production function not only considers capital stock and labour force but also the impact and spill-over of innovation in

$$Q_i = a K_i^{\alpha_1} (L_i + bI_i)^{\alpha_2}$$

with efficiency parameter b and scaling factor a. α_1 and α_2 denote the elasticities of production. The new production factor $D_i = L_i + bI_i$ is called 'know–do'. In the case of $I_i = 0$, the firm i is not innovating and there is no innovation transfer from outside the firm. Then, the production function Q_i becomes the neoclassical production function. Figure 2 [22] illustrates the innovation process in a firm with interaction of labour and capital, know–do, production and profit, and spillovers with industry, scientific system and its own research and development (R&D).

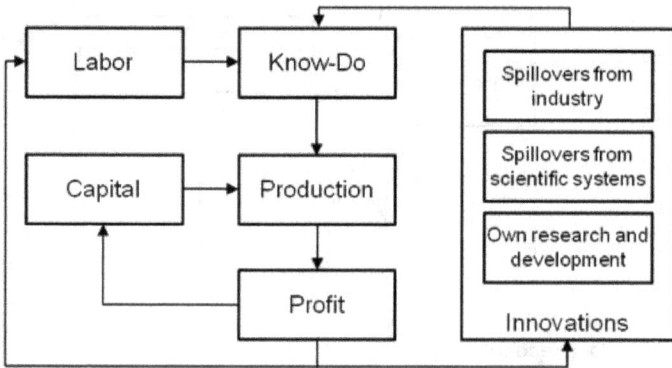

Fig. 2.   Innovation process in a firm.

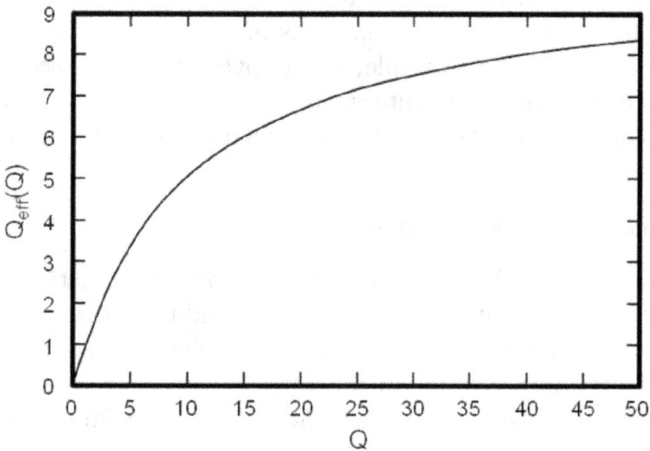

Fig. 3.   Effective production Q_{eff} depending on production Q with maximal production capacity of $Q_{\text{max}} = 10.0$ [23].

A fraction of effective production (Fig. 3) in firm i is used for non-material investment μ_i^K in the capital stock, μ_i^L in the labour force and μ_i^I in innovations with $\mu_i^K + \mu_i^L + \mu_i^I = 1$. The investments in innovations consist of a fixed fraction μ_i^{industry} for own research and development or cooperation with other firms, and μ_i^{science} for cooperation with the scientific system.

We distinguish effective production $Q_{\text{eff},i}$ and a fixed maximal production capacity $Q_{\text{max},i}$ of firm i with respect to its production function Q_i with

$$Q_{\text{eff},i} = \frac{1}{\dfrac{1}{Q_i} + \dfrac{1}{Q_{\text{max},i}}} \quad \text{or} \quad \frac{1}{Q_{\text{eff},i}} = \frac{1}{Q_i} + \frac{1}{Q_{\text{max},i}}.$$

The maximal capacity of production prevents production increasing to arbitrarily high values. In the case of very high values of production Q_i, effective production $Q_{\text{eff},i}$ becomes approximately equal to maximal capacity $Q_{\text{max},i}$. If the production Q_i is much lower than the maximal capacity $Q_{\text{max},i}$, the equation of effective production $Q_{\text{eff},i}$ is equal to the production function Q_i.

Dynamic of capital, labour and innovation

The change in time of the capital used for production by firm i is modelled by the differential equation

$$\frac{dK_i}{dt} = \mu_i^K s_i Q_{\text{eff},i} - \delta_i K_i$$

where $S_i Q_{\text{eff},i}$ is a fraction of the effective production and δ_i is the rate of decrease of the capital stock of firm i.

The change in the labour force over time is given by

$$\frac{dL_i}{dt} = \mu_i^L s_i Q_{\text{eff},i} \frac{1}{w_i n} - \upsilon_i K_i$$

where w_i is the hourly wage rate and n is the number of production periods a worker is employed.

Innovations I_i are measured in terms of working hours spent on the development of new products. The rate of change in innovations is modelled by the differential equation

$$\frac{dI_i}{dt} = \mu_i^I s_i Q_{\text{eff},i} \left(\mu_i^{\text{industry}} \sum_i^N g_{ij} D_j + \mu_i^{\text{science}} \sum_i^M g_{ik}^{\text{science}} D_k^{\text{science}} \right) f(I_i) - \gamma_i$$

where γ_i is the rate of decrease in innovation activities, N, the number of firms and M, the number of scientific institutions. The investments $\mu_i^I Q_{\text{eff},i}$ in innovations are subdivided into a fixed fraction μ_i^{industry} of investments into cooperation with other firms $(i \neq j)$ and own research and development (R&D) $(i = j)$, and another part μ_i^{science} corresponding to the fraction of investment going into cooperation with the scientific system. Thus,

$$\mu_i^{\text{industry}} + \mu_i^{\text{science}} = 1.$$

The know–do D_j is responsible for spill-over effects between firms and the scientific system. The factors g_{ij} and g_{ik}^{science} are interaction coefficients describing the strength of spill-over effects from other systems and the scientific system. For example, a firm which only invests in its own R&D is characterised by $g_{ij} > 0$, $g_{ij} = 0$ and $g_{ik}^{\text{science}} = 0$. For a firm using innovations of other firms without investing into its own R&D, the coefficients are $g_{ii} = 0$, $g_{ij} > 0$ and $g_{ik}^{\text{science}} = 0$.

The spill-over function $f(I_i)$ defines the impact of spill-over effects on firm i depending on its own knowledge stock. By increasing innovation with knowledge-accumulating activities, the spill-over function $f(I_i)$ increases at first, because the lack of knowledge

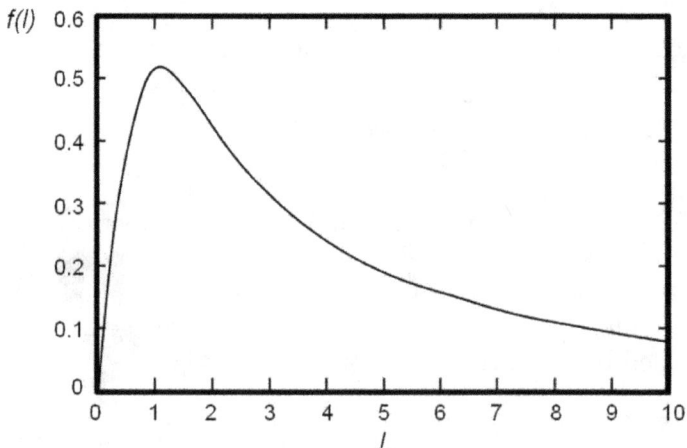

Fig. 4.   Spill-over function $f(I_i)$.

and experience is improved. But, then, with larger values of I_i, it will decrease, because a firm at the progressive frontier has difficulties in extending its knowledge stock through cooperation with others. Therefore, the spill-over curve is assumed to be in the shape of Fig. 4 with the definition

$$f(I_i) = \frac{I_i / I_0}{1 + (I_i / I_0)^2}.$$

Master equation of investment decisions*

Each firm i tries to maximise its profit π_i with respect to its investments in capital K, labour L and innovations I. Therefore, the profit π_i of firm i is defined as returns from effective production $Q_{\text{eff},i}$ reduced by the costs for capital and labour

$$\pi_i = Q_{\text{eff},i} - rK_i - w_i L_i$$

with (constant) wages w_i and the rate of interest r.

The marginal profit $\mu_i^{(m)} = \dfrac{\partial \pi_i}{\partial m}$ is interpreted as utility to invest in capital, labour or innovation with $m = K, L, I$. The probability of changing the type of investment from n to m is determined by the difference $\mu_i^{(m)} - u_i^{(n)} (n \neq m)$. The evolution of the investment ratios $\mu_i^{(m)}$ over time t is modelled by the following master equation in order to include the statistical effects of uncertainty in the decision process:

$$\frac{d\mu_i^{(m)}}{dt} = \varepsilon_i \sum_n f^{(nm)} \mu_i^{(n)} e^{\lambda_i \left(u_i^{(m)} - u_i^{(n)} \right)}$$

$$- \varepsilon_i \sum_n f^{(mn)} \mu_i^{(m)} e^{\lambda_i \left(u_i^{(n)} - u_i^{(m)} \right)}$$

with speed ε_i of adjustment, intensity λ_i of response due to differences of marginal profitabilities and $f^{(nm)}$ barrier effects due to insufficient information between the different investment types. The values of $\mu_i^{(m)}$ range in the interval $0 \leq \mu_i^{(m)} \leq 1$ with the normalisation condition $\sum_m \mu_i^{(m)} = 1$.

The stationary solution with an equilibrium state is given by

$$\mu_{\text{stat}_i}^{(m)} = \frac{e^{2\mu_i^{(m)}}}{\sum_n e^{2\mu_i^{(n)}}} .$$

However, which kind of final state is approached by the investment ratios $\mu_i^{(m)}$, i.e. a stable equilibrium, a limit cycle or chaos, depends on the performance of the firms and other factors.

An example of an agent-based innovation network is a simplified network of two firms which interact with one scientific institution (Fig. 5). The model can be compared with predator–prey interactions of the Lotka–Volterra model. The ratios of investments $\mu_i^{(m)}$ are constant over time. So, the dynamics is studied only with respect to the development of K, L and I.

In the following Table 1 [24], rates of parameters are given with interaction coefficients g_{ij}, g_{ij}^{science}. Their signs + and − indicate positive or negative innovation impulses. The unit of capital stock K_i is measured in scaled currency, and the units of labour force L_i and innovation activity I_i, in scaled working hours.

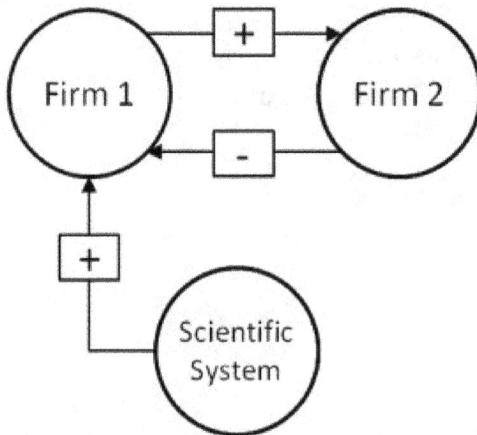

Fig. 5.   Innovation network of two firms and a scientific system.

Table 1.

Global Parameters			Firm-specific Parameters				
			Firm 1	Firm 2		Firm 1	Firm 2
a	1.0	$Q_{max,i}$	10.0	10.0	g_{i1}	0.00	0.78
b	1.0	S_i	0.5	0.5	g_{i2}	−0.60	0.00
α_1	0.5	w_i	0.4	0.4	$g_{i1}^{science}$	Var*	0.0
α_2	0.5	δ_i	0.1	0.1	μ_i^K	0.33	0.33
r	0.1	ν_i	1.0	1.0	μ_i^L	0.33	0.33
$D^{science}$	2.3	γ_i	0.1	0.1	μ_i^I	0.34	0.34
I_0	1.0	–	–	–	$\mu_i^{industry}$	0.5	1.0
–	–	–	–	–	$\mu_i^{science}$	0.5	0.0

Note: *This parameter will vary.

Agent-based innovation dynamics can be illustrated in computer experiments [25]:

In Fig. 6(a), the support from the scientific system is insufficient. The labour force, innovation activity and production decrease towards zero. In Fig. 6(b), the labour force of Firm 2 is increasing rapidly due to positive spill-over effects from Firm 1. A limit cycle appears if the support of the scientific system reaches a critical threshold value (Fig. 6(c)). Both firms are producing profitably, so the labour force and innovation activity for each firm increase and decrease cyclically. At higher values of interaction $g_{11}^{science}$, the trajectories end up at a point of stable equilibrium where both firms can coexist (Fig. 6(d)).

High speed and high capacity of computer simulations allow us to study more complex innovation networks with many interlinked firms and scientific centres. All these simulations demonstrate that large increase in innovation activities can be caused by spill-over effects without random events. Therefore, these results strongly support the strategy to increase know–do and innovation success through the concentration of labour force and knowledge base

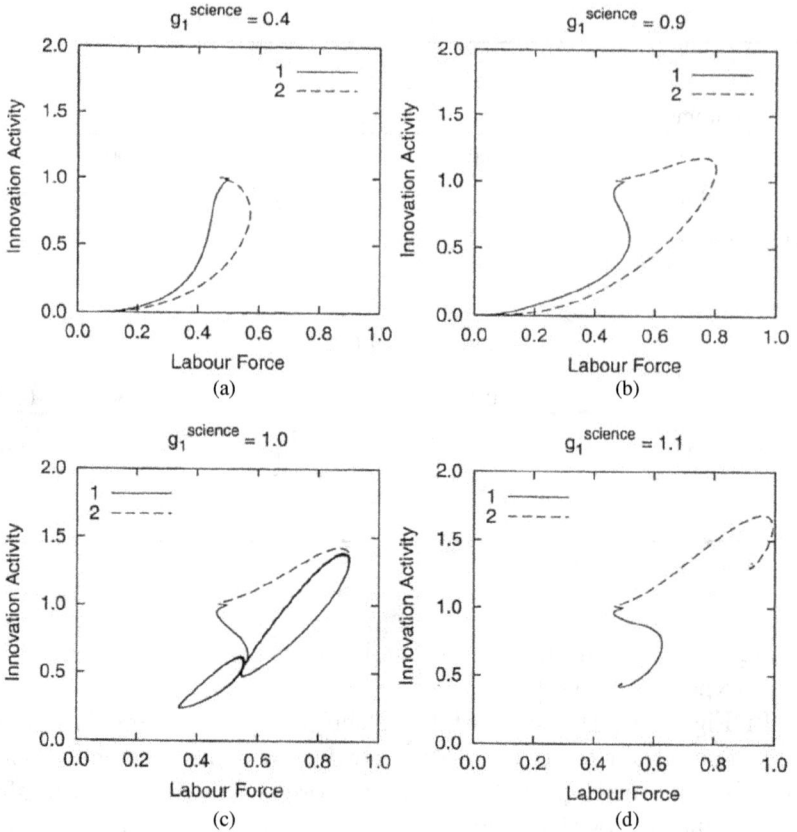

Fig. 6.   Computer simulations of agent-based innovation dynamics.

(i.e. know–do) of several firms and the scientific system. The local activity principle leads to the innovation policy:

> *Drive complex innovation cluster of firms and scientific system at the edge of chaos, in order to be creative!*

Bounded rationality at the edge of chaos

The rationality of human decision is bounded by the wild randomness of markets at the edge of chaos. Human cognitive capabilities are overwhelmed by the complexity of non-linear systems they are forced to manage. Traditional mathematical decision theory assumed perfect rationality of economic agents (homo economicus). Herbert Simon, Nobel Prize laureate of economics and one of the leading pioneers of systems science and artificial intelligence, introduced the principle of bounded rationality in 1959 [26]:

The capacity of the human mind for formulating and solving complex problems is very small compared with the size of the problem whose solution is required for objectively rational behaviour in the real world or even for a reasonable approximation to such objective rationality.

Bounded rationality is not only given by the limitations of human knowledge, information and time, or the incompleteness of our knowledge and the simplification of our model. The constraints of short-term memory and of information storage in long-term memory are well-established. In stressful situations people are overwhelmed by a flood of information, which must be filtered under time pressure. People deviate from game-theoretically predicted equilibria. They act neither in the strict sense of the homo economicus nor completely chaotically. They are locally active agents with non-linear and complex effects of interactions which cannot be predicted in the long run.

Therefore, we must refer to the real features of human information processing and decision making, which are characterised by emotional, subconscious and kinds of affective and non-rational factors. Even experts and managers often prefer to rely on rules of thumb and heuristics, which are based on intuitive feelings of former experience. Experience shows that human intuition not only means lack of information and the failure to make decisions. Our affective behaviour and intuitive feeling are parts of our evolutionary heritage that enable us to make decisions when matters of survival are at stake. Therefore, we must know more about the factual microeconomic action of people, their cognitive and emotional behaviour, in order to understand macroeconomic trends and dynamics. This is the goal of experimental economics, observing, measuring and analysing the behaviour of economic agents with methods of psychology, cognitive and social sciences at, e.g. stock markets or in situations of economic competition.

Human decisions are not only restricted by the complexity of economic markets but also by the complexity of the whole Earth system which is involved in the living conditions of mankind. Environmental and ecological problems can no longer be separated from economic development and consumption of restricted natural resources. It is a challenge of future policies to integrate the economic, ecological and social complexity of our world under the conditions of bounded rationality.

From Earth systems engineering to sustainable innovation

The dynamics of natural systems are increasingly dominated by human influence [27]. In this sense, the Earth can almost be regarded as a human artefact. In fact, there no longer seems to be a natural subsystem on Earth that does not depend on human technology and civilisation. In order to maintain the stability of both natural and human systems, these systems need to be considered in a coupled and co-ordinated way. Research strategies and practical measures are needed to specifically protect and influence these human–nature systems on Earth. This is the goal of Earth systems engineering.

Earth systems engineering and management is a discipline used to analyse, design, engineer and manage complex environmental, economic and technological systems. It not only includes physical, economic and social sciences but also refers to the human factor which is studied in anthropology, ethics and philosophy. Ethics comes in because engineering and managing of the Earth system need responsibility for the whole system. Philosophy claims to surpass disciplinary studies in favour of an integrated view of the whole Earth. Earth systems engineering aims at a rational design and management of coupled human–natural systems in a highly integrated and ethical way [28].

Classical engineering and management demands a high degree of regulation and certainty about the systems behaviour. Therefore, technical systems are strictly separated from the controller, in order to get objective data. But in Earth systems engineering, humans with their cultures and societies are involved in coupled human–natural–technical systems changing and co-evolving together into the future. Adaptive management is a key aspect of Earth systems engineering. It assumes that there is a great deal of uncertainty in environmental systems and holds that there is never a final solution to an Earth systems problem. Therefore, once action has been taken, the Earth systems engineer will need to be in constant interaction with the system, watching for changes and how the system evolves. This way of monitoring and managing ecosystems accepts nature's inherent non-equilibrium dynamics with high risks.

At the outset, a systems analysis process must identify the goals of the system. For this purpose, it is useful to distinguish a descriptive scenario, a normative scenario and a transitive scenario. The descriptive scenario describes the situation as it is and anticipates the future development under these conditions of the status quo [29]. The normative scenario shows the systems development under preferred conditions. From a methodological point of view, the normative scenario is more ambitious, because it must consider the involved stakeholders, monitoring and evaluating the whole system. The transitive scenario changes a system from a descriptive state to a normative state. In the sense of adaptive management, there may be several more or less appropriate solutions. In an iterative and learning process, the variables of the human–environmental–technical system are monitored, changed and adapted, in order to guarantee a self-organising co-evolution of the whole system.

When examining complex human–natural–societal systems, Earth systems engineers must understand how complex systems function. Training in non-linear dynamics will be crucial to fully understand the possible unintended and undesired effects of a proposed Earth systems design. Further on, Earth systems engineers must feel the social, ethical and moral responsibility of the whole Earth System, in order to come up with an objective transitive and normative scenario. Thus, the Earth systems engineer will be expected to explore the ethical implications of the proposed solutions.

In short, contrary to traditional engineering, Earth systems engineers aim at sustainable innovations, considering future generations and the conscious and unconscious effects of the whole system. There is an increasing awareness that the process of development, left to itself, can cause irreversible damage to the environment and the future of the whole system. Therefore, sustainability is an important part to consider in the development of appropriate solutions in Earth systems engineering.

A practical application is industrial ecology. Manufacturing and industrial processes must be transformed from open-loop systems to closed-loop systems. This non-linear feedback system means the recycling of waste to make new products, in order to reduce refuse and increase the effectiveness of resources. The impact of industrial processes on the environment should be minimised.

Earth system engineers create sustainable inventions, concepts and solutions. But entrepreneurship is necessary to realise innovations in economy and society. Thus, the question arises: what does sustainable entrepreneurship mean? The local activities of sustainable entrepreneurs are at the heart of a sustainable future of the whole Earth system, integrating natural capital, economic growth and sustainable innovation.

Sustainability and free markets

Historically, the term 'sustainability' was used to blame economic growth in favour of 'limits to growth' and 'steady state economy'. Actually, limited resources must be wisely developed in concert with nature. But, the evolution of Earth as well as economic development is governed by a non-linear non-equilibrium dynamics which cannot be fixed to a final equilibrium state. Sustainable entrepreneurs have to manage a many-bodies problem and to integrate environmental sustainability, economic sustainability and sociopolitical sustainability. There is a natural limit to resources which must be recognised. But markets are needed for competition and decision of best solutions. Nobody is wise enough to have the master program of the future. Markets mirror the freedom of competing opinions in democracy. All depends on fair boundary conditions in order to find good solutions and to guarantee the welfare of people. Obviously, these mechanisms of democracy and economic markets correspond to the non-linear dynamics of complex systems which also cannot be regulated by a central processor. Appropriate initial and boundary conditions must support a self-organising process aiming at desired goals and intentions. Fair and sustainable conditions of markets refer to social as well as environmental parameters to support the welfare of people depending on the whole Earth system.

The mechanisms of democracy and economic markets correspond to the non-linear dynamics of complex systems with their self-organisation. In contrast, autocratic systems are based on a central processor that must regulate all processes deterministically. The risk of a 'bottleneck' in the flow of information is enormous. In contrast, self-organising processes depend on suitable initial and boundary conditions under which the system finds the desired solutions by itself and with the participation of all system actors. Fair and sustainable market conditions relate to both social and ecological parameters in order to support the well-being of people who are dependent on the entire earth system.

With respect to the whole Earth system, we must also consider the influence of ecological systems. Sometimes, an ecological economics is demanded in order to embed economy into natural evolution according to the laws of thermodynamics and biological systems. Ecological economics intends to improve human well-being through economic development, and designing a sustainable development of ecosystems and societies. Ecological economics is distinguished from neoclassical economics by the assumption that economics is a subfield of ecology. But this is a misunderstanding: Ecology and economy are defined by their own laws which must be coupled and adapted to ensure the well-being of the whole Earth system. Human civilisation has developed its own highly complex rules of interaction which cannot be reduced to biological evolution. Ecology deals with the energy and matter transactions of life and the Earth, while human economy, at least in the past, did not take care of natural resources. The target is to integrate both subsystems in a circular economy without pollution and residuals.

Sustainable entrepreneurship

Sustainable development needs a commitment of business to behave ethically and contribute to economic development while improving the quality of life of the workforce, their families, local communities, the society and the world at large as well as future generations. Sustainable entrepreneurship can be defined as a business venture aiming at sustainable development of the integrated economical, ecological and social system. But how does this relate to the classical meaning of entrepreneurship?

An entrepreneur organises, manages and assumes the risk of a business or enterprise. Often, we use the words 'business' and 'enterprise' interchangeably to refer to the same thing. The word 'entrepreneur' comes from the French word *'entreprendre'*, which means 'to undertake'. In a business context it means to undertake a business venture. But entrepreneurship differs from small business in amount of wealth creation, speed of wealth accumulation, risk and innovation.

Therefore, an entrepreneur is a typical representative of the local activity principle in the sense that an entrepreneur amplifies and transforms investments into wealth, welfare and innovation. Investments are the external supply of an entrepreneurial local activity. But, in a globalised world, these classical goals of entrepreneurship should be realised in a sustainable development with respect to the whole Earth system, i.e. an integrated economic, ecological and social system.

By definition, sustainable entrepreneurs try to realise sustainable business. In any case, sustainable business is an enterprise that has no negative impact on the global or local environment, community, society or economy. Additionally, sustainable businesses have progressive environmental and human rights policies. But business also means profit as driving motivation of entrepreneurs.

Therefore, goals of sustainability should not be misunderstood as brakes to innovation. What is more important is to align the engine of economic profit-making with sustainability goals. The Brundtland Report already emphasised that sustainability has to combine people, planet and profit. Sustainable businesses with the supply chain try to balance all three through sustainable development and sustainable distribution to impact the environment, business growth and the society. Sustainable development within a business can create value for customers, investors and the environment. A sustainable business must meet customer needs while, at the same time, treating the environment well.

These ambitious goals can only be realised by sustainable technology and innovation. Companies must focus on their ability to change their products and services towards less waste production and sustainable best practices. Innovation needs cooperation and the formation of networks with partner companies and the scientific system with synergetic effects and fruitful spillovers. Continuous process surveying and improvement is essential to reduction in waste. Employee awareness of the company-wide sustainability plan further aids the integration of new and improved processes. Periodic monitoring and reporting of the company's performance is necessary to follow the message of sustainable entrepreneurship.

However, these demands must not degenerate into an excessive bureaucracy of regulations that paralyse innovation and competitiveness of a company. Competitiveness means the profitability, effectiveness and viability of a company. These requirements for a successful company have not changed since Adam Smith.

References

[1] Mandelbrot, B. B. and Hudson, R. L. (2004), *The (mis)Behaviour of Markets. A Fractal View of Risk, Ruin, and Reward*, Basic Books: New York.

[2] Weidlich, W. (2002), *Sociodynamics. A Systematic Approach to Mathematical Modelling in the Social Sciences*, Taylor & Francis: London.

[3] Helbing, D. (2008), *Managing Complexity: Insights, Concepts, Applications*, Springer: Berlin.

[4] Mantegna, R. N. and Stanley, H. E. (2000), *An Introduction to Econophysics. Correlations and Complexity in Finance*, Cambridge University Press: Cambridge.

[5] McCauley, J. L. (2004), *Dynamics of Markets. Econophysics and Finance*, Cambridge University Press: Cambridge.

[6] Arthur, W. B., Durlauf, S. N., and Lane, D. A. (Ed.) (1997), The economy as an evolving complex system II, in *Proceedings Volume of the Santa Fé Institute*, Bd. XXVII, Reading Mass.

[7] Ebeling, W., Freund, J., and Schweitzer, F. (1998), *Komplexe Strukturen: Entropie und Information*, B. G. Teubner: Leipzig.

[8] Mainzer, K. (2009), Challenges of complexity in economics, *Evolutionary and Institutional Economics Review*, 6(1), pp. 1–22. Japan Association for Evolutionary Economics.

[9] Mainzer, K. (2007), *Der kreative Zufall. Wie das Neue in die Welt kommt*, C. H. Beck: München (also Japanese translation 2011).

[10] Haken, H. (1990), *Synergetics. An Introduction*, 3rd edn., Springer: Berlin.

[11] Mainzer, K. and Chua, L. (2013), *Local Activity Principle*, Imperial College Press: London.

[12] Weidlich, W. and Braun, M. (1992), The master equation approach to nonlinear economics, *Journal of Evolutionary Economics*, 2, pp. 233–265.

[13] Mainzer, K. (2014), *Die Berechnung der Welt. Von der Weltformel zu Big Data*, C. H. Beck: München.

[14] Mainzer, K. (2016), *Information. Algorithmus — Wahrscheinlichkeit — Komplexität — Quantenwelt — Leben — Gehirn — Gesellschaft*, Berlin University Press: Berlin.

[15] Mainzer, K. (2010), Challenges of complexity in management, in Schloemer, S. and Tomaschek, N. (Eds.) *Leading in Complexity. New Ways of Management*, Heidelberg: Carl-Auer: Heidelberg, pp. 5–23.

[16] Schumpeter, J. A. (1911), *Theorie der wirtschaftlichen Entwicklung. Eine Untersuchung über Unternehmergewinn, Kapital, Kredit, Zins und den Konjunkturzyklus*, 9th edn., Duncker & Humblot (1997). English translation: *The Theory of Economic Development. An Inquiry Into Profits, Capital, Credit, Interest, and the Business Cycle*, Harvard University Press: Cambridge (1934).

[17] Fischer, M. M. and Fröhlich, J. (2001), *Knowledge, Complexity, and Innovations Systems*, Springer: Berlin.

[18] Christensen, C. M. (1997), *The Innovator's Dilemma. When New Technologies Cause Great Firms to Fail*, Harvard Business School Press: Boston, MA.

[19] Lakatos, I. and Musgrave, A. (Ed.) (1970), *Criticism and the Growth of Knowledge*, Cambridge University Press: London.

[20] Haag, G. and Liedl, P. (2002), Modelling of knowledge, capital formation, and innovation behaviour within micro-based profit oriented and correlated decision processes, in *Fischer, Fröhlich 2001* (Anmerkung 17), pp. 251–273.

[21] Haag, G. and Liedl, P. (2002), Modelling of knowledge, capital formation, and innovation behaviour within micro-based profit oriented and correlated decision processes, in *Fischer, Fröhlich 2001* (Anmerkung 17), p. 255.

[22] Haag, G. and Liedl, P. (2002), Modelling of knowledge, capital formation, and innovation behaviour within micro-based profit oriented and correlated decision processes, in *Fischer, Fröhlich 2001* (Anmerkung 17), p. 257.

[23] Haag, G. and Liedl, P. (2002), Modelling of knowledge, capital formation, and innovation behaviour within micro-based profit oriented and correlated decision processes, in *Fischer, Fröhlich* 2001 (Anmerkung 17), p. 238.

[24] Haag, G. and Liedl, P. (2002), Modelling of knowledge, capital formation, and innovation behaviour within micro-based profit oriented and correlated decision processes, in *Fischer, Fröhlich 2001* (Anmerkung 17), p. 270.

[25] Haag, G. and Liedl, P. (2002), Modelling of knowledge, capital formation, and innovation behaviour within micro-based profit oriented and correlated decision processes, in *Fischer, Fröhlich 2001* (Anmerkung 17), p. 266.

[26] Simon, H. (1957), *Administrative Behavior: A Study of Decision-making Processes in Administrative Organizations*, Macmillian: New York, p. 198.

[27] Allenby, B. R. (2005), *Reconstructing Earth. Technology and Environment in the Age of Humans*, Island Press: Washington, DC.

[28] Newton, L. H. (2003), *Ethics and Sustainability. Sustainable Development and the Moral Life*, Prentice Hall: Upper Saddle River, N.J.

[29] Gibson, J. E. and Scherer, W. T. (2007), *How to do Systems Analysis*, J. Wiley & Sons: Hoboken, New Jersey.

Chapter 4

Innovation of Artificial Intelligence

Information and communication networks are the background to global automation through artificial intelligence. In the future, enormous computer capacities will be needed to cope with the vast amounts of data generated by this civilisation. The complexity of life, its misunderstood interrelationships, its sensitivity and vulnerability, which is evident in diseases such as cancer as well as viral pandemics, requires new tools in life sciences and medicine. Bioinformatics will increasingly have to rely on machine learning and suitable computer and storage capacities. The same applies to overcoming global financial and economic crises that require early warning systems. Information and communication networks are thus growing together with the other major high-tech hype, artificial intelligence.

The digitalisation of human civilisation requires an enormous amount of energy. Computer technology and information and communication networks should therefore not be seen separately from the energy consumption of this civilisation. This also highlights the ecological dimension of digitalisation, as fossil fuels pollute the environment. What can we learn from nature, which has produced extremely effective and energy-saving brains and nervous systems in the course of evolution? This is where neuromorphic computing on memristive architectures and photonic computers comes in, whose computer architectures are modelled on the brains of evolution. Instead of the energy-guzzling von Neumann architectures of classic computers from smartphones and PCs to supercomputers, computer units are now being used that still work on hardware but according to the efficient methods of neurons and synapses in brains. Instead of total digitalisation, they also use the advantages of analogue processes, as known from living organisms.

After electrification and digitalisation in the 20th century, the quantisation of communication and supply networks is now imminent. This will happen gradually and not 'disruptively'. The universal quantum computer will also not 'disruptively' replace classic computer technology, but will increasingly be embedded in classic and neuromorphic computer structures and solve new tasks that were impossible with these processes. There is already talk of hybrid computer networks or digital ecologies that are gradually spreading across the world.

Digital AI systems

Definition of artificial intelligence

Traditionally, AI (Artificial Intelligence) has been understood as a simulation of intelligent human thinking and acting. This definition suffers from the fact that 'intelligent human thinking' and 'acting' are not defined. Furthermore, human is made the yardstick of intelligence, although evolution has produced many organisms with varying degrees of 'intelligence'. In addition, we have long been surrounded in technology by 'intelligent' systems which, although they are independent and efficient, are often different from humans in controlling our civilisation. All the more the question arises as to what makes us human.

Einstein has answered the question 'What is time?' as a physicist based on measurement results independent of humans: 'Time is what a clock measures'. Therefore, we propose a working definition that is independent of human beings and only depends on measurable quantities of systems. To this end, we look at systems that can solve problems more or less independently. Examples of such systems could be organisms, brains, robots, automobiles, smartphones or accessories that we wear on our bodies (wearables). Systems with varying degrees of intelligence are also available at factory facilities (Industry 4.0), transport systems or energy systems (smart grids) which control themselves more or less independently and solve central supply problems. The degree of intelligence of such systems depends on the degree of self-reliance, the complexity of the problem to be solved and the efficiency of the problem-solving procedure.

So, there is no 'the' intelligence but degrees of intelligence. Complexity and efficiency are measurable variables in computer science and engineering. An autonomous vehicle then has a degree of intelligence that depends on its ability to reach a specified destination independently and efficiently. There are already more or less autonomous vehicles. The degree of their independence is technically precisely defined. The ability of our smartphones to communicate with us is also changing. In any case, our working definition of intelligent systems covers the research that has been working successfully for many years in computer science and technology under the title 'Artificial Intelligence' and is developing intelligent systems [1].

> A system is called intelligent when it can solve problems independently and efficiently. The degree of intelligence depends on the degree of autonomy of the system, the degree of complexity of the problem and the degree of efficiency of the problem-solving procedure. The list of the mentioned criteria is not complete, but it can be extended in the sense of a 'living' definition if necessary.

It is true that intelligent technical systems, even if they have a high degree of independent and efficient capacity for problem-solving, were ultimately initiated by people. But even human intelligence has not fallen from the heaven and depends on specifications and limitations. The human organism is a product of evolution that is full of molecularly and neuronally encoded algorithms. Humans have developed over millions of years and are only more or less efficient. Randomness often played along. This has resulted in a hybrid system of abilities that by no means represents 'the' intelligence at all. AI and technology have long since overtaken natural skills or solved them differently. Think of the speed of data processing or storage capacities. There was no such thing as 'consciousness' as necessary for humans. Evolutionary organisms such as stick insects, wolves or humans solve their problems in different ways. In addition, intelligence in nature is by no means dependent on individual organisms. The swarm intelligence of an animal population is created by the interaction of many organisms, similar to the intelligent infrastructures that already surround us in technology and society. Even in this application we can distinguish degrees of autonomous decision making and acting.

Symbolic AI: Logic and deduction

On the basis of digital computability, AI was orientated in a first phase towards the formal (symbolic) calculus of logic, with which solutions to problems can be derived in a rule-based manner. This is why we also speak of symbolic AI. A typical example is an automatic proof with logical deductions, which can be realised with computer programs. Automation also means autonomy to a certain degree, as computer programs take over the proof activities of a mathematician. Knowledge-based expert systems are computer programs that store and accumulate knowledge about a specific field, automatically drawing conclusions from the knowledge to offer solutions to concrete problems in the field. Unlike human experts, however, the knowledge of an expert system is limited to a specialised information base without general and structural knowledge about the world [2, 3].

To build an expert system, the expert's knowledge must be put into rules, translated into a program language and processed with a problem-solving strategy. The architecture of an expert system therefore consists of the following components: Knowledge base, problem-solving component (derivation system), explanation component, knowledge acquisition, and dialogue component. In this architecture, the limits of symbolic AI become clear at the same time: abilities that cannot or can only with difficulty be captured symbolically and simulated rule-based behaviour remain closed to symbolic AI.

Subsymbolic AI: Statistics and induction

Sensory and motor skills are not logically derived from textbook knowledge, but are learned, trained and practised from examples. In this way, we learn to move motorically and to recognise patterns and connections in a multitude of sensory data, which we can use to orient our actions and decisions. Since these abilities do not depend on their symbolic representation, we also speak of subsymbolic AI. The formal conclusions of logic are now replaced by the statistics of the data. In statistical learning, general dependencies and correlations are to be deduced from finitely many observational data by algorithms [4]. Deduction in symbolic AI is thus replaced by induction in subsymbolic AI. To this end, we can imagine a natural science experiment in which a series of changed conditions (inputs) are followed by corresponding results (outputs). In medicine, it could be a patient who reacts to medication in a certain way.

Here we assume that the corresponding pairs of input and output data are generated independently by the same random experiment. Statistically, we therefore say that the finite sequence of observation data $(x_1, y_1), \dots, (x_n, y_n)$ with inputs x_i and outputs $y_i (i = 1, \dots, n$ is realised by random variables $(X_1, Y_1), \dots, (X_n, Y_n)$, which are based on a probability distribution $P_{X,Y}$. Algorithms should now derive properties of the probability distribution $P_{X,Y}$. An example would be the expected probability with which a corresponding output occurs for a given input. However, it can also be a classification task: a set of data is to be divided into two classes. What is the probability that an element of the dataset (input) belongs to one class or the other (output)? In this case, we also speak of binary pattern recognition.

The current successes of machine learning seem to confirm the thesis that what matters is the largest possible datasets, which are processed with ever stronger computer power. However, the recognised regularities then only depend on the probability distribution of the statistical data.

Statistical learning attempts to derive a probabilistic model from finitely many data of outcomes (e.g. random experiments) and observations.

Statistical reasoning, conversely, attempts to infer properties of observed data from an assumed statistical model.

Learning with neuronal networks

Neural networks with learning algorithms play a key role in the automation of statistical learning (Chapter 9). Neural networks are simplified computational models modelled on the human brain in which neurons are connected with synapses. The intensities of the neurochemical signals sent between the neurons are represented in the model by numbers as synapse weights. Probabilistic networks experimentally bear a strong resemblance to biological neural networks. If cells are removed or individual synapse weights are changed by small amounts, they prove to be fault-tolerant to minor disturbances like the human brain, for example, in the case of minor accidental damage. The human brain works with layers of parallel signal processing. For example, between a sensory input layer and a motor output layer, there are internal intermediate steps of neuronal signal processing that are not connected to the outside world.

In fact, the representation and problem-solving capacity can also be increased in technical neural networks by interposing different layers with as many neurons as possible that are capable of learning. The first layer receives the input pattern. Each neuron of this layer has connections to each neuron of the next layer. Activation functions ensure the transmission of signals and activate the downstream neurons ('firing'). The back-to-back connection continues until the last layer is reached and emits a pattern of activity [5].

We speak of supervised learning procedures when the prototype to be learned (e.g. the recognition of a pattern) is known and the respective error deviations can be measured against it. A learning algorithm must change the synaptic weights until a pattern of activity in the output layer emerges that deviates as little as possible from the prototype.

An effective method is to calculate the error deviation of actual and desired outputs for each neuron in the output layer and then backpropagate it through the layers of the network. We then speak of a backpropagation algorithm. The intention is to reduce the error to zero or negligibly small values through a sufficient number of learning steps for a default pattern.

As civilisation becomes increasingly dependent on digitalisation and artificial intelligence, considerable security risks are emerging. Nobody understands or can control in detail what happens, for example, in the non-linear interactions of the neurons and synapses of a neural network in machine learning. This is why these neural networks are 'black boxes' for users and developers, raising fundamental questions of security, trust in technology and responsibility.

Artificial neural networks are extremely effective in dealing with complex problems (real-world problems). What is missing, however, are specifications and standards for the security of their outputs. To achieve this, the black box of neural networks must be better understood, controlled and verified. However, the verification of neural networks is a tough cognitive problem: even the proof of simple properties proves to be NP-complete within the framework of complexity theory. The reasons for this are the size of the networks used in practice (scaling) and the non-linear activation functions of their neurons, which cannot be reproduced by humans to this extent and at this speed.

As neural networks are also subject to the dynamics of complex systems, they are often sensitive to small disturbances and changes in their inputs, which can build up to uncontrollable effects. The robustness and stability of the networks is therefore closely related to their security.

Different verifications can be specified for different classes of neural networks, which can be derived from various theories of logic and mathematics [6]. These include verification methods based on the satisfiability of Boolean propositional logic formulae (SAT = Satisfiability Theories) [7], satisfiability of first-level predicate logic formulae (SMT = Satisfiability Modulo Theories) [8], reduction to linear problems (MIP = Mixed Integer Linear Programming) [9] and the robustness of multi-layer perceptron networks (MLP = Multi-Layer Perceptron). In SAT and SMT verifications, the classical AI (symbolic AI) of automated reasoning is combined with machine learning [10]. MIP is based on the logic and algebra of linear programming. Robustness analyses of MLP apply findings from the theory of complex dynamic systems in machine learning.

Neural networks as complex dynamical systems

Neural networks are the key to the success of machine learning in technical applications and commercial practice. However, networks with hundreds and thousands of layers and millions of neurons are now being used. Depth refers to the number of neuronal layers. We no longer just talk about deep learning, but also about 'deepest learning' with this enormous number of layers. The training and processes of these networks can only be described statistically and appear random. Verifications of their behaviour are generally only based on selective tests with trial-and-error procedures. This exacerbates the black box problem of these networks with their growing depth and breadth in practice.

Is there a mathematical theory to understand and explain the processes in these large neural networks? In fact, neural networks can be explained in the mathematical theory of complex dynamic systems, which has been the guiding principle in this book (and in my work for many years). The historical model was statistical mechanics, in which, for example, a fluid is understood as a complex system consisting of many fluid molecules as elements. Macroscopic behaviours such as flow patterns are derived from the non-linear interactions of the microscopic elements. However, the mathematical formalism used is independent of the application and can therefore be used for modelling in the natural, economic and social sciences [11]. It is therefore not a 'naturalistic' methodology but a mathematical methodology that is independent of application models.

Artificial neurons are used as microscopic elements of deep neural networks. Their interactions are represented by synaptic weights. Macroscopic behaviour patterns are described by statistical distribution functions. The emergence of macroscopic patterns and structures takes place in phase transitions that depend on critical control parameters.

Figure 1 shows a multi-layered neural network with input x, which is transformed into the output $f(x; \theta)$ with a sequence of intermediate signals $s^{(1)}$, $s^{(2)}$ and $s^{(3)}$ of intermediate layers. The parameters θ represent the numerical values with which the signals in the synapses between the neurons are weighted depending on their 'firing' thresholds. In practical and commercial applications, this can involve 100 billion parameters. The parameters θ are the control parameters of the complex system.

Leaving aside the layers, the parameterised function $f(x; \theta)$ represents the neural network as a complex system with input x and control parameter θ. Due to the enormous number of parameters, in practice

$$f(x; \theta)$$

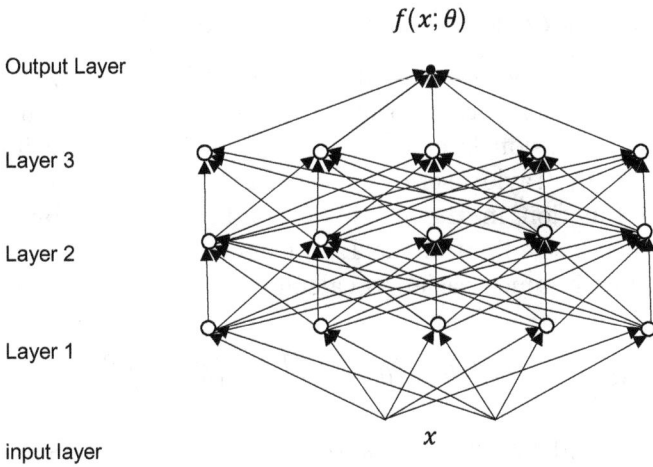

Output Layer

Layer 3

Layer 2

Layer 1

input layer

$$x$$

Fig. 1. Multi-layered neural network as a complex dynamical system.

the network is parameterised with a random selection of a probability distribution $p(\theta)$. In a next step $\theta \to \theta^*$ the parameters are adjusted in such a way that the resulting network function $f(x; \theta^*)$ approximates a desired target function $f(x)$ as close as possible with $f(x; \theta^*) \approx f(x)$.

In order to find θ^*, the network function $f(x; \theta)$ is trained with data $(x, f(x))$, which were observed in the target function. Obviously, it is an inductive learning process which is basic for machine learning and which was already explained in the beginning of section 'Digital AI systems'. The adjusting of parameters is called 'training' and its procedure, 'training algorithm'.

In the sense of complex dynamic systems, the macroscopic behaviour of the trained network function $f(x; \theta^*)$ must be explained from the microscopic descriptions of the network with the trained parameter values θ^*. To understand the target approximation $f(x; \theta^*) \approx f(x)$, it is necessary to explain how the trained network $f(x; \theta^*)$ utilises the training data $(x, f(x))$ in the approximation of $f(x)$. The practical challenge lies in mastering the huge number of parameters.

Computation of complex neural networks*

Mathematically, the network is represented by a highly non-linear function $f(x; \theta^*)$. In order to make it mathematically manageable, its Taylor expansion is examined in the vicinity of the initialised value of θ. A Taylor series is used in calculus to represent a (smooth) function in the neighbourhood of a point by a power series, which is the limit of the Taylor polynomials. Polynomials are mathematically much easier to calculate than other functions [12]:

$$f\left(x; \theta^*\right) = f\left(x; \theta\right) + \left(\theta^* - \theta\right)\frac{df}{d\theta} + \frac{1}{2}\left(\theta^* - \theta\right)^2 \frac{d^2 f}{d\theta^2} + \cdots$$

with $f(x; \theta)$ and its derivations for the initialised values of parameter θ.

But there are fundamental computational problems for the Taylor series of network function $f(x; \theta^*)$:

(1) An infinite number of terms $f, \frac{df}{d\theta}, \frac{d^2 f}{d\theta^2}, \ldots$ must be computed. As the differences $(\theta^* - \theta)$ between the trained and initialised parameters become large, the number of needed terms increases for an appropriate approximation of the trained network function $f(x; \theta^*)$.

(2) As the parameters θ are a random sample of the initialised distribution $p(\theta)$, one gets a different function $f(x; \theta)$ with each initialisation of the network. Each term $f, \frac{df}{d\theta}, \frac{d^2 f}{d\theta^2}, \ldots$ becomes a random function of the input x. Each initialisation leads to a distribution over the network function and its derivations, which are represented in a mapping

$$p\left(\theta\right) \rightarrow p\left(f, \frac{df}{d\theta}, \frac{d^2 f}{d\theta^2}, \cdots\right)$$

of the probability distribution of the initial parameters θ on the joint probability distribution of the network function $f(x; \theta)$, it's gradient $\frac{df}{d\theta}$, it's Hesse matrix $\frac{d^2 f}{d\theta^2}$, and so on. This joint probability distribution collects an infinite number of random functions which are analytically not tractable.

(3) The learnt value of parameter θ^* is the result of a complex train-ing process. The learning algorithms work iteratively in learning steps which non-linearly depend on each other in a complex way and which are practically hardly tractable.

If the three problems (1–3) could be solved, the Taylor develop-ment would be an appropriate explanation of the trained network function $f(x; \theta^*)$. In this case, one would have to find a probability distribution

$$p\left(f^*\right) \equiv p(f\left(x; \theta^*\right)|\text{learning algorithm, training data})$$

over the trained network functions $f(x; \theta^*)$ in dependence of the learning algorithm and the training data.

Deep and deepest learning depend on the size of the networks, which is determined through the depth of the network, i.e. the number L of layers, and its width n, in other words the number of neurons in a layer. In order to simplify the distribution $p(f^*)$ over trained networks, it is obvious to analyse the behaviour of extremely large networks in theoretical limits. Therefore, idealised networks with infinite width

$$\lim_{n \to \infty} p\left(f^*\right)$$

are assumed. In this case, one gets extreme simplifications of the computation:

(1) All higher derivative terms $\frac{d^k f}{d\theta^k}$ mit $k \geq 2$ disappear and only the linear terms $f, \frac{df}{d\theta}$ remain.
(2) The distributions of random functions are independent and lead to a very simple expression

$$\lim_{n \to \infty} p\left(f, \frac{df}{d\theta}, \frac{d^2 f}{d\theta^2}, \ldots\right) = p(f)\, p\left(\frac{df}{d\theta}\right).$$

(3) The training process is linear and completely independent of all details of the learning algorithm. Therefore, one can find an analytic solution of θ^*. In this case, the trained distribution

$\lim\limits_{n\to\infty} p(f^*)$ is a simple Gauß distribution (with a mean different from zero).

Unfortunately, such a formal boundary value approach with the limit of infinite network width only allows very simple models of deep neural networks. All detailed correlations and precise details of neurons that play a role in real, very large but finite networks are eliminated.

Therefore, it is proposed in the following to correct the limit for networks with infinite width in such a way that the corrections become smaller as the width increases. To this end, ideas from perturbation theory, which is used in physics for interacting systems, are taken up. The process of deep learning is now described by a $1/n$ expansion with the reciprocal value $\epsilon = 1/n$ of the layer width n and a small expansion parameter $\epsilon \ll 1$. Mathematically, the limit value with infinite width is supplemented by a Taylor expansion with the following correction terms [13]:

$$p(f^*) \equiv p^{(0)}(f^*) + \frac{p^{(1)}(f^*)}{n} + \frac{p^{(2)}(f^*)}{n^2} + \cdots$$

with $p^{(0)}\left(f^*\right) \equiv \lim\limits_{n\to\infty} p(f^*)$ and corrections $p^{(k)}\left(f^*\right)$ with $k \geq 1$. It can be shown that even a calculation of the first correction term makes the problems (1)–(3) of a Taylor expansion manageable. In this case, the Taylor expansion is truncated except for the first two terms:

$$p\left(f^*\right) \equiv p^{(0)}\left(f^*\right) + \frac{p^{(1)}\left(f^*\right)}{n} + O\left(\frac{1}{n^2}\right).$$

(1) Higher derivative terms $\frac{d^k f}{d\theta^k}$ with $k \geq 4$ contribute values in the order of magnitude $1/n^2$ and smaller. To consider the essential contributions in the order of magnitude $1/n$, the four terms $f, \frac{df}{d\theta}, \frac{d^2 f}{d\theta^2}, \frac{d^3 f}{d\theta^3}$ are sufficient.

(2) The distribution of the random functions for initialisation $p\left(f, \frac{df}{d\theta}, \frac{d^2f}{d\theta^2}, \frac{d^3f}{d\theta^3}\right)$ is almost simple in the order of magnitude and can be handled in perturbation theory in all details.

(3) In this case, the non-linear training dynamics is tractable with perturbation theory and an analytical solution of θ^* can be found.

The trained distribution of the order of magnitude $1/n$ corresponds to a fast Gaussian distribution that can be analysed. The restriction to the order of $1/n$ is sufficient to calculate and understand the distribution $p(f^*)$ over the trained network functions. Details of the interactions between the neurons are now taken into account. This approach therefore corresponds much more closely to real technical neural networks than the approach with networks of infinite width.

The ratio of width n and depth L of a neural network allows a classification of the mode of action of layered networks. The following applies to the proportion $r \equiv L/n$:

- In the limit $r \to 0$, the interactions between neurons are eliminated. Networks with limit $\lim_{n\to\infty} p(f^*)$ of infinite width are actually not really deep, as the relative depth with $\frac{L}{n} = 0$ is zero.
- There are significant interactions between neurons in the interval $0 < r \ll 1$. The restriction to networks of finite width of the order of $1/n$ allows precise calculations of the output of the trained network. In this sense, such networks are effectively deep.
- In the interval $r \gg 1$, the neurons are strongly connected. Then they behave chaotically. Due to the strong fluctuations, effective calculation is no longer possible. In this sense, these networks are 'too deep'.

For practical and commercial applications, it follows that large networks of finite width promise the greatest success in terms of effective computability and controllability. It is remarkable that this central insight for practical applications follows from an abstract mathematical theory in which the analysis of the infinite plays a central role. Once again, it is confirmed that a good mathematical theory is essential in order to understand and explain a problem in depth.

From statistical to causal learning

Statistical learning and inference from data are therefore not enough. Rather, we need to recognise the causal relationships of causes and effects behind the measurement data [14]. These causal connections depend on the laws of the respective application domain of our research methods, i.e. the laws of physics, the laws of biochemistry and cell growth in the example of cancer research, etc. If it were otherwise, we could already solve the problems of this world with the methods of statistical learning and reasoning.

> Statistical learning and reasoning without causal domain knowledge is blind — no matter how large the amount of data (Big Data) and computing power!

The debate between probabilistic and causal thinking is by no means new but was already fought out epistemologically in 18th century philosophy between David Hume (1711–1776) and Immanuel Kant (1724–1804). According to Hume, all cognition is based on sensual impressions (data) that are psychologically 'associated'. According to this, there are no causality laws of cause and effect but only associations of impressions (e.g. lightning and thunder) that are 'habitually' correlated with (statistical) frequency [15]. According to Kant, laws of causality are possible as rationally formed hypotheses that can be tested experimentally. Their formation is not based on psychological associations but on the category of causality [16], which can be assumed and verified for predictions based on experience. According to Kant, this procedure has been in use in physics since Galileo Galilei, which thus first became a science through causal explanations and not only statistical correlations.

Thus, in addition to the statistics of the data, additional law and structure assumptions of the application domains are needed, which are verified by experiments and interventions. Causal explanatory models (e.g. the planetary model or a tumour model) fulfil the law and structure assumptions of a theory (e.g. Newton's theory of gravity or the laws of cell biology).

In causal reasoning, properties of data and observations are derived from causal models, i.e. law assumptions of causes and effects. Causal reasoning thus makes it possible to determine the effects of interventions or data changes (e.g. through experiments).

Conversely, causal learning attempts to derive a causal model from observations, measurement data and interventions (e.g. experiments) that presuppose additional law and structural assumptions.

A highly topical technical example of the growing complexity of neural networks is a self-learning vehicle. For example, a simple automobile with various sensors (e.g. neighbourhood, light, collision) and motor equipment can already generate complex behaviour through a self-organising neural network. If neighbouring sensors are excited in a collision with an external object, then so are the neurons of a corresponding neural network connected to the sensors. This creates a wiring pattern in the neural network that represents the external object. In principle, this process is similar to the perception of an external object by an organism — only much more complex there.

If we now imagine that this automobile will be equipped with a 'memory' (database) with which it can remember such dangerous collisions in order to avoid them in the future, then we have an inkling of how the automobile industry will be on its way to building self-learning vehicles in the future. They will be very different from conventional driver-assistance systems with pre-programmed behaviour under certain conditions. It will be neuronal learning as we know it in nature from more highly developed organisms.

But how many real accidents are needed to train self-learning ('autonomous') vehicles? Who is responsible when autonomous vehicles are involved in accidents? What ethical and legal challenges arise? In complex systems such as neural networks with, for example, millions of elements and billions of synaptic connections, the laws of statistical physics do allow us to make global statements about trend and convergence behaviour of the entire system. However, the number of empirical parameters of the individual elements may be so large that no local causes can be identified. The neural network remains a 'black box' for us. From an engineering point of view, authors therefore speak of a 'dark secret' at the heart of machine learning AI: '... even the engineers who designed [the machine learning-based system] may struggle to isolate the reason for any single action' [17].

Two different approaches in software engineering are conceivable:

1. testing only shows (randomly) found errors but not all other possible ones;
2. to avoid them in principle, a formal verification of the neural network and its underlying causal processes would have to be carried out.

In summary, machine learning with digital neural networks works but we cannot understand and control the neural network processes in detail. Today's machine learning techniques are mostly based only on statistical learning, but that is not enough for safety-critical systems. Therefore, machine learning should be combined with proof assistants and causal learning. Correct behaviour is guaranteed by metatheorems in a logical formalism [18].

Towards hybrid AI

This model of self-learning vehicles is reminiscent of the organisation of learning in the human organism, where behaviour and reactions also take place largely unconsciously. 'Unconscious' means that we are not aware of the causal processes of the locomotor system, which is controlled by sensory and neuronal signals. This can be automated using statistical learning algorithms. However, this is not enough in critical situations. In order to achieve greater safety through better control in the human organism, the mind must intervene with causal analysis and logical reasoning. This process should be automated in machine learning using causal learning algorithms and logical reasoning assistants.

> The goal is therefore a hybrid AI in which symbolic and subsymbolic AI are combined in a manner analogous to the human organism.

Robots as AI systems

With the increasing complexity and automation of technology, robots are becoming service providers for industrial society. The evolution of living organisms today inspires the construction of robotic systems for different purposes. As the complexity and difficulty of the service task increases, the use of AI technology becomes unavoidable. And robots don't have to look like humans. Just as airplanes do not look like birds, there are also other adapted shapes depending on their function. So, the question arises: for what purpose should humanoid robots possess which properties and abilities?

Humanoid robots should be able to act directly in the human environment. In the human environment, the environment is adapted to human proportions. The design ranges from the width of the corridors and the height of a stair step to the positions of door handles. For non-human robots (e.g. on wheels and with other grippers instead of hands), large investments for environmental changes would have to be made. In addition, all tools that humans and robots should use together are adapted to human needs. Not to be underestimated is the experience that humanoid forms psychologically facilitate the emotional handling of robots.

Humanoid robots

But humanoid robots don't just have two legs and two arms. They also have optical and acoustic sensors. In terms of space and battery life, there have so far been limitations on the processors and sensors that can be used. Miniaturisation of optical and acoustic functions is just as necessary as the development of distributed microprocessors for local signal processing. The aim of humanoid robotics is for humanoid robots to be able to move freely in a normal environment, overcome stairs and obstacles, search for paths independently, remain mobile after a fall, operate doors independently and carry out work while leaning on one arm. In principle, a humanoid robot could then walk like a human.

To reach the final stage of cohabitation with humans, robots must be able to visualise humans in order to become sufficiently sensitive. This requires cognitive abilities. A distinction can be made between the three stages of the functionalist, connectionist and action-oriented approaches, which will now be analysed [20].

The basic assumption of functionalism is that there is an internal cognitive structure in living beings, such as robots, which represents objects in the external world with their properties, relationships and functions using symbols.

This is also referred to as functionalism, as the processes of the external world are assumed to be isomorphically mapped in functions of a symbolic model. Similar to how a geometric vector or state space represents the motion sequences of physics, such models would represent the environment of a robot. The functionalist approach goes back to the early cognitivist psychology of the 1950s by A. Newell and H. Simon [21]. The symbols are processed in a formal language (e.g. computer program) according to rules that establish logical relationships between the representations of the outside world, allowing conclusions to be drawn and thus allowing knowledge to be generated.

According to the cognitivistic approach, rule processing is independent of a biological organism or robot body. In principle, all higher cognitive abilities such as object recognition, image interpretation, problem-solving, speech comprehension and awareness could be reduced to symbolic calculation processes. Consequently, biological abilities like consciousness should be transferable to technical systems.

The cognitivist–functionalist approach has proved its worth for limited applications, but it has fundamental practical and theoretical limitations. A robot of this kind requires a complete symbolic representation of the outside world, which must be constantly adjusted as the robot's position changes. Relations such as ON (TABLE, BALL), ON (TABLE, CUP), BEHIND (CUP, BALL), etc., which represent the relation of a ball and a cup on a table relative to a robot, change as the robot moves around the table.

Humans, on the other hand, do not need symbolic representation and no symbolic updating of changing situations. They interact sensorially–physically with their environment. Rational thoughts with internal symbolic representation do not guarantee rational action, as simple everyday situations already show. Thus, we avoid a sudden traffic obstruction due to lightning-fast physical signals and interactions, without resorting to symbolic representations and logical derivations. This is where subsymbolic AI comes into play.

In cognitive science, we therefore distinguish between formal and physical actions [22]. Chess is a formal game with complete symbolic representation, precise game positions and formal operations. Soccer is a non-formal game with skills that depend on physical interactions without full representation of situations and operations. There are rules to the game. But because of the physical action, situations are never exactly identical and therefore cannot be reproduced at will (in contrast to chess).

> The connectionist approach therefore emphasises that meaning is not carried by symbols, but results from the interaction between different communicating units of a complex network. This formation or emergence of meanings and patterns of action is made possible by the self-organising dynamics of neural networks [23].

However, both the cognitivistic and the connectionistic approaches can, in principle, disregard the environment of the systems and only describe symbolic representation or neuronal dynamics.

> In contrast, the action-oriented approach focuses on embedding the robot body in its environment. In particular, simple organisms of nature such as bacteria suggest to build behaviour-controlled artefacts that are able to adapt to changing environments.

However, here too the demand would be one-sided, to favour only behaviour-based robotics and to exclude symbolic representations and models of the world. It is true to say that human cognitive performance takes into account functionalist, connectionist and behavioural aspects.

It is therefore correct to assume, as in humans, a humanoid robot's own corporeality (embodiment). Then these machines operate with their robot body in a physical environment and establish a causal relationship to it. They each have their own experiences with their bodies in this environment and should be able to build their own internal symbolic representations and systems of meaning [24].

How can such robots independently assess changing situations? Physical experiences of the robot begin with perceptions about sensor data of the environment. They are stored in a relational database of the robot as its memory. The relations of the outside world objects form causal networks with each other, on which the robot orients its actions. In this context, a distinction is made between events, persons, places, situations and objects of daily use. Possible scenarios and situations are described with propositions of a formal logic.

Cyberphysical systems as AI systems

In evolution, intelligent behaviour is by no means limited to individual organisms. Sociobiology regards populations as superorganisms capable of collective performance. The corresponding abilities are often not completely programmed in the individual organisms and cannot be realised by them alone. An example is the swarm intelligence of insects, which can be seen in termite structures and ant trails [25]. Human societies with extrasomatic information storage and communication systems are also developing collective intelligence which only shows itself in their institutions.

Collective patterns and cluster formations can also be observed in populations of simple robots without having been programmed beforehand. Robot populations as service providers could find concrete application in road traffic with driverless transport systems or forklifts, which communicate independently about their behaviour in certain traffic and order situations. Increasingly, different robot types such as driving and flying robots (e.g. for military missions or space exploration) will interact with each other [26].

R. A. Brooks of the Massachusetts Institute of Technology generally calls for a behavioural AI which is based on artificial social intelligence in robot populations [27]. Social interaction and coordination of common actions in changing situations is an extremely successful form of intelligence that has evolved over evolution. Even simple robots, like simple organisms of evolution, could generate collective achievements. In management, one speaks of social intelligence as a soft skill that should now also be considered by robot populations.

Autonomous reactions in different situations without human intervention are a major challenge for AI research. Decision-making algorithms are best improved in real road traffic. Similarly, a human driver improves their skills through driving experience.

Self-driving vehicles or robot cars are cars that can drive, steer and park without a human driver.

Highly automated driving lies between assisted driving, in which the driver is supported by driver assistance systems, and autonomous driving, in which the vehicle drives itself without the driver's intervention.

In highly automated driving, the vehicle only partially has its own intelligence, which plans ahead and could take over the driving task, at least in most situations. Man and machine work together.

Traditional computer systems were characterised by a strict separation of the physical and virtual worlds. Mechatronic control systems, which are installed in modern vehicles and aeroplanes, for example, and consist of a large number of sensors and actuators, no longer correspond to this image. These systems recognise their physical environment, process this information and can also influence the physical environment in a coordinated manner. The next stage in the development of mechatronic systems is 'cyberphysical systems' (CPS), which are not only characterised by a strong link between the physical application model and the computer control model but are also embedded in the working and everyday environment (e.g. integrated intelligent energy supply systems) [28, 29]. Due to their networked embedding in system environments, CPS go beyond isolated mechatronic systems.

CPS consist of many networked components that coordinate with each other independently for a common task. They are therefore more than the sum of the many different small smart devices in ubiquitous computing, as they realise overall systems consisting of many intelligent subsystems with integrating functions for specific goals and tasks (e.g. efficient energy supply). This extends intelligent functions from the individual subsystems to the external environment of the overall system. Like the Internet, CPS are becoming collective social systems, but in addition to information flows, they also integrate energy, material and metabolic flows (like mechatronic systems and organisms).

Industry 4.0 alludes to the previous phases of industrialisation. Industry 1.0 was the age of the steam engine. Industry 2.0 was Henry Ford's assembly line. The assembly line is nothing more than an algorithmisation of the work process, which produces a product step by step according to a fixed programme through the division of labour between people. In Industry 3.0, industrial robots intervene in the production process. However, they are fixed in place and always work through the same programme for a specific subtask. In Industry 4.0, the work process is integrated into the Internet of Things. Workpieces communicate with each other, with transport equipment and with the people involved in order to organise the work process flexibly.

Trust in artificial intelligence

AI is often portrayed as a threat to human labour. However, the coronavirus crisis also showed how AI and robotics could step in when humans fail to keep the economy running, how digital communication and healthcare could be supported and how the solution can be found in a learning process together with human intelligence, for example, in the form of a vaccine. After corona, it cannot be ruled out that we will be hit by even more dangerous pandemics. For the future, it would therefore be desirable if learning AI could be used to simulate possible changes in viruses in advance in order to develop a toolbox for the rapid composition of vaccines — with AI algorithms produced in advance, so to speak.

In order to promote trust in AI tools, they must be certified like all technical tools. We are working on such 'DIN standards' in a steering group for an AI roadmap on behalf of the German government. Ultimately, AI should be a service for us humans. We therefore also need to strengthen human judgement and value orientation so that algorithms and big data do not get out of hand.

Recent dramatic accidents illustrate the dangers of software errors and system failures in safety-critical systems. Program errors and system failures can lead to disasters: in medicine, massive overdoses caused by the software of a radiotherapy machine in 1985–1987 caused the death of patients in some cases. In 1996, the explosion of the Ariane 5 rocket caused a sensation due to a software error. The most recent example is the software error and system failure of the Boeing 737 max. Verification tests are traditionally an integral part of program development in software engineering. Once the requirements, design and implementation of a computer program have been determined, it is usually verified and then predictive maintenance is carried out for the duration of its use in order to initiate replacement and repair before a machine part fails due to wear and tear, for example.

A computer program is called correctness or certification if it can be verified that it follows a given specification. In practice, verification procedures with varying degrees of accuracy and therefore reliability are used [30]. However, for reasons of time, effort and cost, many users are content with random sample tests only. Ideally, however, a computer program should be as reliable as a mathematical proof. To this end, proof programs ('proof assistants') have been developed with which a computer program can be checked for correctness automatically or interactively with a user.

The idea originally comes from the mathematical proof theory of the early 20th century, when important logicians and mathematicians such as David Hilbert, Kurt Gödel and Gerhard Gentzen formalised mathematical theories in order to then prove, for example, the correctness, completeness or freedom from contradiction of these formalisms (and thus of the mathematical theories in question). The formalisms are now computer programs. Their proofs of correctness must themselves be constructive in order to exclude any doubt about their certainty. Proof assistants are being investigated at both Munich Universities in Germany, the LMU (Ludwig Maximilians University) and the TUM (Technical University of Munich) [31]. The French proof assistant Coq goes back to the French logician and mathematician Thierry Coquand, among others, and its name is reminiscent of the French heraldic animal [32, 33].

This shows very clearly how current questions about the security of modern software and AI are rooted in fundamental questions of logic and philosophy. A central question is how modern machine learning can be controlled by such proof assistants [34]. Ultimately, the challenge is whether and how AI programs can be certified before they are unleashed on humanity. Statistical learning as it is practised today often works in practice, but the causal processes often remain misunderstood and a black box. Statistical testing and trial and error is not enough for safety-critical systems. In the future, it will therefore be important to combine causal learning with certified AI programs using proof assistants, even if this may seem complex and ambitious for practitioners.

AI systems as service systems

AI programs are now found not only in individual robots and computers. Algorithms capable of learning are already controlling the processes of a networked world with exponentially growing computing capacity. Without them, the flood of data on the Internet generated by billions of sensors and networked devices would be unmanageable. Thanks to the sensors, things now also communicate with each other and not just people. This is why we talk about the Internet of Things (IoT).

In medicine and the healthcare system, large hospital centres are examples of such complex infrastructures whose coordination of patients, doctors, medical staff, technical devices, robotics and other service providers would no longer be controllable without IT and AI support.

The safety-critical challenges that have just been discussed will become even more acute in such infrastructures. Beyond this, however, there is the question of the role of humans in a more or less automated world. I am therefore in favour of technology design that goes beyond technology assessment. Experience has shown that the traditional view of simply letting developers get on with their work and assessing the consequences of their results at the end is not enough. In the end, the child may have fallen into the well and it is too late. Innovation cannot be planned. But we can incentivise the desired results. Ethics would then not be a brake on innovation, but an incentive for desired innovation. Such an ethical, legal, social and ecological roadmap of technology design for AI systems would correspond to the basic idea of the social market economy, according to which there is room for manoeuvring for competition and innovation. The benchmark remains the dignity of the individual, as laid down in the Fundamental Law of the constitution as the highest axiom of parliamentary democracy.

This ethical positioning in the global competition for AI technology is by no means a matter of course. For the global IT and AI companies of Silicon Valley, it is ultimately about a successful business model, even if they promote IT infrastructures in less developed countries under conditions that they dictate. The other global competitor, however, is China, which strictly follows a state monopoly in the Silk Road project. The Chinese Social Core project is closely linked to the ambitious goal of producing the world's fastest supercomputers and most powerful AI programs. This is the only way to realise the Social Core with the total data collection of all citizens and their central evaluation. The total state control of private data may shock European observers, but it is accepted by a large majority of Chinese people. One reason for this is the greater efficiency in solving global threats such as epidemics. This includes direct access to all kinds of medical data for medical research. In addition, there is another value tradition that has been practised in China for centuries: in this country's Confucian tradition, the highest standard of value is collective harmony and security rather than the autonomy of the individual with enforceable rights to freedom.

The proclamation of individual human rights is deeply rooted in the philosophical tradition of European democracies such as Kant's doctrine of categorical imperative [35]. We do need certified AI algorithms as a reliable service for coping with the complexity of civilisation. However, it is also crucial to strengthen human judgement and value orientation so that algorithms and big data do not get out of hand. In the global competition between AI systems, we should be able to shape our own living environment according to our own values.

Generative AI systems

What does generative AI mean?

A spectacular application example of subsymbolic AI are chatbots like ChatGPT, which, because of its amazing capabilities as an automatic text generator, had more followers than social media such as Instagram and Spotify within a few days, with millions of users, since it was launched in 30 November 2022. ChatGPT can generate texts from school assignments at grammar school level to texts of seminar papers of middle university level. Based on a 'large language model' (LLM), this AI program can be used to talk about business plans or to commission the writing of a song, poem or novel fragments in a certain style.

In fact, ChatGPT's language model is based on a massive amount of text (Big Data) that has been trained into the system by humans. It is thus an example of machine learning based on statistical learning theory and pattern recognition, as explained in the previous section. The ambitious goal here is to overcome a key limitation of symbolic AI, which in its knowledge-based expert systems was limited to the expertise of specialists (e.g. medical expertise in a specific medical discipline), provided it could be translated into logical rule-based formulae. With the increase in computing power and the handling of large masses of data with models of statistical learning, the goal is now being pursued to also bring the general 'world knowledge' of us humans to the machine.

For this purpose, the chatbot is trained with texts from news, books, social media, online forums, images, films and spoken language texts. Algorithms are used to learn from the training data. The chatbot reproduces patterns that it recognises in the stored data. This is done using the same procedures that are used in face recognition to recognise images of people from image files. The reproduced texts are compared with trained sample texts and thus gradually improved by reinforcement learning algorithms. Corrections can also be made if correlations of the trained data lead to discrimination, for example. Similar to indoctrinated humans, such misbehaviour can never be ruled out due to the volume of the trained datasets. Since these chatbots are widely accepted in social media, they can also cause dangerous disinformation.

Ultimately, ChatGPT is also nothing more than a stochastic machine that recombines and reconfigures data, texts, images and spoken words with pattern recognition algorithms. However, due to modern computer technologies that can store enormous amounts of data and apply fast learning algorithms, amazing results are produced that simulate a great deal of human background knowledge and intuition. But this also reveals the mechanisms on which our conversational and cultural worlds are based — reproductions and recombinations of patterns that can largely be adopted by machines. Even the social sciences, cultural studies and the humanities are not immune to this, not to mention journalism.

Wittgenstein called these 'language games' that function according to certain linguistic rules. The original often consists only in a small change and variation of the usual language games and 'narratives'. In machine learning, there is now talk of 'stochastic parrots'. Positively speaking, ChatGPT is therefore suitable for exposing the mechanisms of the culture industry and journalism. They will have to become more sophisticated in order not to be replaced by machines.

In our book *Limits of AI* [36], the central weakness of statistical learning theory and machine learning was highlighted in contrast to mathematical and logical thinking. ChatGPT can also write and evaluate computer programs only by imitating and recombining stored templates and fragments — but at an astonishingly high level that cannot be distinguished even by 'educated' humans. The difference to human thinking is already demonstrated by a gifted pupil: without having been 'fed' with all kinds of textbooks, he or she solves a mathematics problem without the effort and memory volume of a chatbot.

In the 'machine room' of ChatGPT

Technically, ChatGPT is an LLM that generates human-like texts with deep learning algorithms from large data masses of speech. It is based on a 'Generative Pre-trained Transformer' (GPT) architecture, in which a transformer generates texts with a neural network. The model is trained beforehand with large data masses of books, articles, web pages, etc. to recognise patterns and structures of natural languages. Given an input (called 'prompt'), the model generates a suitable text based on the previously trained knowledge.

By using a transformer, the GPT differs from previous linguistic models that sequentially predicted probable words in a text context. Transformers process all input data simultaneously. Fundamental to this is a process of 'self-attention', which distributes changing weights for different parts of the input data with reference to other positions in the speech sequence. Due to increasing computational efficiency, the GPT models have been extended and improved since 2018 from GPT1 to GPT4 for ever larger and more diverse scopes of knowledge.

A self-attention method uses a neural network to weight the importance of different parts of the input and make predictions. The input is mapped to multiple keys, values and queries that correspond to learned weight matrices. The model then calculates the scalar product of the queries with the keys for all items of the input. This produces a score for each item. These scores are then used to calculate a weight ('attention') for each item in the input. The scores are multiplied by these attention weights to add up these products as the output of the self-attention process. This output is now connected to the input and passes through the multiple layers of the feedforward neural network that realises self-attention.

To better match the outputs of the ChatGPT with the user's intentions, a reinforcement learning from human feedback (RLHF) algorithm is used, which distinguishes the following three steps [37]:

Step 1: Supervised Fine-Tuning Model
The first development involved fine-tuning the GPT-3 model by hiring 40 contractors to create a supervised training dataset, in which the input has a known output for the model to learn from. Inputs, or prompts, were collected from actual user entries into the Open API (Application Programming Interface). The labellers then wrote an appropriate response to the prompt thus creating a known output for each input. The GPT-3 model was then fine-tuned using this new, supervised dataset, to create GPT-3.5, also called the supervised fine-tuning (SFT) model.

In short: step 1 collects demonstration data and trains a supervised policy with the following partial steps:

- A prompt is sampled from the prompt dataset. The prompt dataset is a series of prompts previously submitted to the open API.
- A labeller demonstrates the desired output behaviour. 40 contractors were hired to write responses to prompts.
- These data are used to fine-tune GPT-3 with supervised learning. Input–output pairs are used to train a supervised model on appropriate responses to instructions.

Step 2: Reward Model

After the SFT model is trained in step 1, the model generates better aligned responses to user prompts. The next refinement comes in the form of training a reward model (RM) in which a model input is a series of prompts and responses, and the output is a scalar value, called a reward. The reward model is realised by reinforcement learning in which a model learns to produce outputs to maximise its reward in step 3.

In short: step 2 collects comparison data and trains a reward model in the following partial steps:

- A prompt and several model outputs are sampled. Responses are generated by the SFT model.
- A labeller ranks the outputs from best to worst.
- These data are used to train our reward model. Combinations of rankings are served to the model as a batch datapoint.

In order to speed up comparison collection, labellers with responses of rankings were used. It delivers comparisons for each prompt shown to a labeller. Comparisons are correlated within each labelling task.

Step 3: Reinforcement Learning Model

In the final stage, the model is presented with a random prompt and returns a response. The response is generated using the policy that the model has learnt in step 2. The policy represents a strategy that the machine has learnt to use to achieve its goal of maximising its reward. Based on the reward model developed in step 2, a scalar reward value is then determined for the prompt and response pair. The reward then feeds back into the model to evolve the policy.

In short: step 3 optimises a policy against the reward model using reinforcement learning with Proximal Policy Optimisation (PPO). PPO is a policy gradient method for reinforcement learning which alternates between sampling data through interaction with the environment and optimising an objective function using stochastic ascent. Whereas standard policy gradient methods perform one gradient update per data sample, PPO proposes a novel objective function that enables multiple epochs of minibatch updates:

- A new prompt is sampled from the dataset.
- The policy generates an output. A policy is a strategy that an agent uses in pursuit of goals.
- The reward model calculates a reward for the output.
- The reward is used to update the policy using PPO. (Kullback–Leibler penalty for SFT model is used to avoid overfitting.)

Challenges of ChatGPT for societal policies

The analyses in the previous sections show that the chatbot ChatGPT is not magic, but is based on computable algorithms of stochastics and statistical learning theory. Therefore, its performance and limitations can also be clearly assessed. Neither euphoria nor excessive timidity are hence appropriate to the matter. ChatGPT has caused unease especially in everyday life, media, communication, economy, healthcare, and so on. The question lurks everywhere whether professions in these fields could be replaced by chatbots in the future. Against the background of the foundational analysis of chatbots, the following will assess the significance of ChatGPT for concrete job profiles in societal application.

Example: ChatGPT as healthcare expert

AI research is opening up new methods of prevention, diagnosis and therapy in medicine, from assisted early detection of diseases to personalised treatments [38]. Current AI methods in medicine focus on the areas of knowledge-based systems, pattern analysis and pattern recognition as well as robotics. Classification systems support diagnostic imaging procedures in areas such as laboratory diagnostics, parasitology, radiology, pathology, cytology, dermatology and ophthalmology, as well as in minimally invasive surgery.

Knowledge-based systems date back to the beginnings of AI, when algorithms were used to derive solutions to problems (e.g. diagnosis) from a knowledge base (e.g. data and symptoms of diseases). In this sense, a specialised activity of a medical expert was simulated in a limited area of application. We therefore also speak of medical expert systems. From the point of view of computer science, the derivation of a diagnosis, for example, is realised by an algorithm, which is also referred to as an inference engine. The inference engine is programmed by entering knowledge.

The knowledge is represented declaratively. It consists of factual knowledge, as in a database, or rule knowledge in the form of symbolic production rules, according to which certain activities are to be carried out under certain conditions. As in symbolic logic, knowledge and problem-solving are represented by derivations in computer programs. Knowledge-based systems in medicine are therefore an example of symbolic AI.

Modern basic medical research is essentially based on molecular biological knowledge, which would no longer be possible without the support of bioinformatics. Bioinformatics is an interdisciplinary field of research that combines biology, computer science, mathematics and statistics with engineering. The focus is on algorithms and software that are used to analyse complex structures and functions of, for example, proteins from molecular data. This involves statistical pattern analysis and pattern recognition in huge amounts of data that can no longer be logically derived from a few premises. For this purpose, molecules must be represented in codes. The search methods for corresponding problem solutions and codes increasingly rely on machine learning methods, which are no longer orientated towards formal logic but towards statistical learning theory.

With ChatGPT, AI has also arrived in the everyday medical routine of a doctor's practice or hospital. In principle, all text-based communication and data analysis can be taken over by medical chatbots that are trained using corresponding datasets. This may indeed lead to greater efficiency in patient communication between medical assistants. But how do medical diagnoses work with ChatGPT? Can we rely on diagnoses from a chatbot that we also use for other purposes in everyday life? This is where the first limits become apparent.

The language of medicine is extremely complex and standardised. What seems like a plausible and well-formulated answer to the lay person can be wrong, skewed and misleading. Similar to 'solving maths problems' in mathematics, 'solving medical cases' is therefore a central field of training at university for students of medicine. Here, ChatGPT can be used in a didactically meaningful way by critically analysing and discussing the chatbot's answers in the seminar or in exercise groups by the students in order to improve their own problem-solving skills. The chatbot can also help to write linguistic summaries of complex medical cases for specific purposes. But especially in the medical field, it is ultimately a matter of responsibility up to and including liability, which cannot be delegated to an automaton.

It is to be expected that medical chatbots will become better and better, trained by comprehensive datasets. Which doctor will then dare to give a different diagnosis to the suggestion of a highly specialised chatbot that has access to all the medical databases currently available for this application? Added to this is the time pressure in a doctor's surgery with many patient files to decide on.

As is generally the case with today's AI, Achilles' heel of modern machine learning is also evident here: all learning results depend on data extraction and therefore the quality of the chatbots' training data, which can be deliberately or unconsciously manipulated, erroneous and problematic. The huge amount of data used as a basis is not the only decisive factor. In the end, medical judgement is indispensable in order to make responsible diagnoses — no matter how highly specialised machine learning is.

Example: ChatGPT as psychotherapist

It becomes particularly sensitive in professions of mental health where language is used to convey feelings and empathy, such as psychologists and psychotherapists. Weizenbaum's early language program ELIZA was already intended to simulate a psychotherapist. At the time, Weizenbaum was appalled at how this simple program was accepted as a psychotherapeutic interlocutor: people projected their own desires and fears into this program. With ChatGPT, automated interlocutors become conceivable that can be used as substitutes for human interlocutors. This could be an extremely problematic business model for a psychotherapist who uses chatbots en masse in order to collect fees for such conversations. This would not only be profit-seeking but extremely dangerous for psychologically vulnerable patients. Tests show that the chatbot also generates false and skewed information. The chatbot could be used as a transcriber of conversations or for advice based on available data, which would then have to be critically proofread. In all these applications, it must be clear that the chatbot only performs statistical data analysis based on large data masses with pattern recognition. It can therefore only understand and convey feelings and empathy to the extent that previously trained texts spoke about them. In training, answers from the chatbot can be critically assessed by students in order to train their own psycho-therapeutic judgement.

Example: ChatGPT as personnel manager

For entry into a profession, personnel managers play a central role in the various companies. They assess the suitability of applicants on the basis of written documents and personal interviews. In the process, a standardisation of questions can be observed, to which desired standard answers can be given. However, a standardised assessment procedure can easily be simulated with the current services of chatbots. Standard questions must therefore be avoided. Interactions in the assessment must play a stronger role than written surveys according to standard questionnaires. In the end, human resource management is also not about text generation, but decisions. However, there will be dips and changes in personnel marketing. A job advertisement or careers website can be easily and professionally written by ChatGPT.

Example: ChatGPT as programmer

ChatGPT already writes simple programs in computer science. In fact, the programming profession and the systems architect profession can be expected to change without being replaced by AI. Indeed, ChatGPT can already provide (simple) building blocks of programming to be used in writing more complex programs. At the same time, however, this will make the programming profession more demanding and professional. It should also be taken into account that the neural networks of chatbots will only be one example of programs that will change the work of programmers in the future. However, program verification will be all the more important in the future. The smallest errors in elementary building blocks that are 'automatically' generated by chatbots such as

ChatGPT can have a catastrophic effect on the entire software if they are not recognised in time. Therefore, the high qualification of programmers is indispensable.

Example: ChatGPT as journalist

In the media and journalists' associations, ChatGPT is sometimes perceived as a threat. In fact, this chatbot writes desired articles and essays in perfect national languages. Routine articles could definitely be done automatically. If the journalist wishes, the linguistic style could also be adapted to a particular writer. So, in the sense of the Turing test, these writers are replaceable. Bans on the chatbot, which are demanded by some professional associations, are of little help here. Rather, one must learn to deal with this technology and improve one's own performance. For more demanding texts, ChatGPT could help pre-structure and incorporate the necessary data. The editor should exercise control and responsibility (also in the legal sense). In particular, false information that would otherwise be reproduced and passed on by the AI should be weeded out. One could also distribute chatbots in the network, which 'spontaneously' make statements in the desired context and pass on propaganda and disinformation. So, the challenges in the media sector are great but so are the opportunities for improving quality. In journalism training, the chatbot could generate sample articles on certain topics, which are then critically assessed by the students in order to improve their later work.

Example: ChatGPT as lawyer

Language-dependent professions are also legal professionals as, for example, lawyers, prosecutors or judges. Thus, it is conceivable to entrust ChatGPT with the task of a business lawyer. A company wishes to have articles of association for a certain legal form of a company (e.g. in Germany, a GmbH). For this, a company describes its profile by answering certain standard questions. The chatbot then automatically drafts the company's articles of association. Legal databases already exist, but they generate many answers and options to queries, which a lawyer must laboriously work through. Since public prosecutors and judges, for example, suffer from the enormous flood of pending cases and trials, they could all be too happy to rely on the quick and seemingly efficient help of a chatbot.

The same applies to legal issues in the healthcare sector, such as health insurance. Can we rely on the advice of ChatGPT? Will we end up communicating with a highly specialised legal chatbot rather than a clerk at the health insurance company when it comes to questions of care and pensions? Who bears the ultimate responsibility for such automated decisions? And that could be really dangerous.

The reasons are obvious: law in particular shows the clear limits of today's chatbots. The language of law is extremely complex and standardised. What seems like a plausible and well-formulated answer to the lay person can be wrong, skewed and misleading. Similar to 'solving maths problems' in mathematics, 'solving legal cases' is therefore a central field of training at university for students of law. Here, too, ChatGPT can be used in a didactically meaningful way by critically analysing and discussing the chatbot's answers in the seminar or in exercise groups by the students in order to improve their own problem-solving skills. The chatbot can also help to write linguistic summaries of complex judgements and legal cases for specific purposes. But especially in the legal field, it is ultimately a matter of responsibility up to and including liability, which cannot be delegated to an automaton.

Potential and limitation of ChatGPT

The possibilities of ChatGPT at school and university should therefore not lead to bans but to the critical question: are examinations, as we traditionally know them, still up-to-date and appropriate in a changed working world with different technical conditions? Therefore, first of all, a fundamental discussion is required that asks about the possibilities and limits of this technology. We need to know the algorithms in order to be able to assess the possibilities and limits. This requires basic theoretical knowledge and also practical experience in dealing with these programs. So, learners should first be given a basic understanding of machine learning and the special algorithms of chatbots like ChatGPT. Then comes their own experimentation with orders to the chatbot and the evaluation of its answers.

It is important to understand, for example, that this is reinforcement learning, in which new and modified answers are given through constant questioning, which in the best case improve. However, this depends on the chatbot's knowledge base, which was previously taught to the chatbot through supervised learning in a training phase with a human supervisor. It follows that an initial response from the chatbot is not yet directly usable, but requires post-processing and correction. In this iterated way, sample solutions in the various disciplines could be generated in dialogue with the chatbot.

An appropriate use of ChatGPT in examinations therefore depends on the boundary conditions [39]. Only in oral examinations and written examinations can the examiner largely ensure that there is no cheating. However, it is also a question of scale whether in some subjects hundreds or even thousands of candidates have to be examined or a manageable small number. For written assignments, how sure one can be depends on the subject. In fact, it becomes difficult with text generation tasks in the cultural sciences and humanities. Empirical papers in the social sciences and economics are based on empirical data that can be controlled by checking sources.

Incidentally, in the natural sciences, for example, it is quite conceivable that in the case of specialist articles the linguistic formulations in the terminology typical for the subject or the structuring of the article in the manner typical for the subject could be generated by a chatbot while the results of the actual new laboratory discovery only have to be inserted. Accordingly, there are examination performances in which verbalisations only make up a part of the examination performance. This refers to laboratory experiments, statistical analyses or programming. It must be admitted, however, that the chatbot undermines the ability to argue and present thoughts in writing and oral presentation. These skills are, for example, quite central for leadership tasks in a company. Exams that no longer reflect these skills are therefore of little help to a company. Here, the limits of chatbots must be critically evaluated and other forms of examinations, such as examination interviews, must be demanded [40].

From subject to subject, it must be examined exactly how the respective subject competence can be replaced by a chatbot. It must not be forgotten that examination performance is also an important tool for students to recognise their own abilities, talents and limitations in order to find a suitable career later on. The objectivity of examinations must remain an important yardstick for awarding scholarships and university positions, for example. From a legal point of view, it must therefore be ensured that the misuse of technical aids such as ChatGPT can be established with legal certainty. Already under the impression of the pandemic a legal order was passed in Bavaria to be able to take electronic distance examinations. Accordingly, a legal framework must be created to regulate the use of chatbots such as ChatGPT in university examinations. Questions of equal opportunities and compliance with data protection standards will play a crucial role.

DeepSeek — The competitor from China

The next step in generative AI after ChatGPT is the Chinese start-up DeepSeek. With minimal use of resources, its AI model R1 successfully competes against tech giants such as OpenAI and Google [41]. DeepSeek is a generative AI platform that, like all machine learning algorithms, is trained to recognise complex data patterns in unstructured datasets. Following the announcement of this technology, the shares of graphics chip manufacturer Nvidia fell by almost 17%. This corresponds to a loss of USD 589 billion in stock market value. DeepSeek's app became the most downloaded app in the Apple App Store, pushing the AI giant ChatGPT from OpenAI into second place.

A glance into the machine room of DeepSeek

Unlike other AI systems, DeepSeek has a certain degree of autonomous learning capabilities. The platform improves its capabilities from new data, experience and feedback. DeepSeek's versatility is striking as it can be applied to text, image, audio and sensor data in equal measure. In practical terms, it enables both medical image analyses and climate research with complex patterns in weather and environmental data. Data sources from these different areas can be integrated and analysed in real time. With fast processing speed, it is not only possible to optimise processes but also to make surprising discoveries of deeper insights.

In contrast to the ChatGPT model with SFT, DeepSeek R1 uses reinforcement learning to evaluate different possible solutions. This enables a deeper analysis of tasks, which leads to complex conclusions. This development of conclusions therefore takes into account re-evaluations of intermediate steps and thus achieves quality improvement through self-correction. When solving a maths problem, for example, the algorithm receives feedback (rewards) at each step as to how good or bad the previously chosen solution path is.

Complex tasks in particular are not solved by the algorithm randomly starting with one aspect and then trawling through all the possibilities. This would not only be inefficient but also extremely energy-intensive. Instead, resources are first used to determine how a problem can be better solved and divided into subtasks. A clever human problem solver would proceed in a similar way and first think about a suitable solution strategy. As with human problem-solving, 'aha' moments can occur to re-evaluate an intermediate step, make corrections and analyse the problem at a deeper level. In DeepSeek, these 'aha' moments are characterised by corresponding tokens.

For this purpose, all steps that are necessary for a successful solution must be described in a large language model. This method is called Chain-of-Thought (CoT) which was already used by OpenAI. The Chain-of-Thought prompting simulates reasoning processes by dividing complex tasks in a sequence of logical steps up to a final solution.

From prompt-chaining to CoT prompting

Prompt chaining is a simplified precursor to CoT prompting [42]. The program has to generate an answer to a question based on a specific context. In contrast, CoT prompting goes beyond simply answering. Rather, CoT prompting aims to create a comprehensive and logically consistent argument [43].

For example, a chatbot would generate the answer 'Mars is red' to the question 'What colour is Mars?'. In contrast, CoT prompting explains why Mars appears red, i.e. which materials on the surface of this planet cause light refractions that appear red. Another path of explanation could show the cultural–historical and mythological significance of the talk of the 'red planet of the god of war'. Chain-of-Thought prompting requires LLMs that can be used to generate reasoning steps that can be applied to similar tasks. This improves the generative system itself. Mathematical text tasks can also be comprehensively understood and solved with all possible variants.

In the meantime, CoT prompting has been further developed into different variants that are tailored to different application domains and extend and improve the reasoning capabilities of the model. The zero-shot variant of CoT prompting uses the knowledge already contained in models to solve problems without prior specific training data. For example, when asked which politician from a particular party is the interior minister of a particular country, Zero-Shot CoT draws on its existing knowledge of a country's political parties to generate an answer without having been trained on this application domain.

The automatic Chain-of-Thought (Auto-CoT) prompting aims to minimise the manual effort involved in creating prompts by auto-mating the generation and selection of effective thought paths. This variant extends the application of CoT to a wider range of tasks. The multimodal thought chain extends the CoT program to include input from different application areas with texts and images. In this way, different types of information can be integrated for complex thinking tasks.

CoT prompting thus proves to be a technique to improve the performance of LLMs in complex reasoning tasks. Some advantages are as follows:

- *Improved prompt outputs*: CoT prompting improves the efficiency of LLMs for complex logical tasks, which are divided into more simple steps.
- *Transparency and understanding*: The generation of intermediate steps explains how a model can derive its conclusions. It makes the decision process more understandable.
- *Multi-layered argumentation*: CoT prompting leads to more accurate answers for tasks which need multi-layered ('deeper') thinking. This cognitive ability is needed for solving complex problems, making decisions and understanding cause–effect connections.

CoT prompting also has the following restrictions:

- *High computational power*: Generating and processing multiple reasoning steps requires more computing power and time compared to the standard one-step prompt. Adopting this technique is therefore associated with higher costs.
- *Expensive and labour-intensive*: The development of effective CoT prompts can be more complex and labour-intensive and requires a deep understanding of the problem domain and abilities of the model.

Innovations in prompt engineering have significantly improved the understanding of models and their interaction with the original prompt. This enables more differentiated and contextualised reasoning. In addition, logical reasoning is also integrated into statistical-only machine learning approaches.

The integration into symbolic and logical reasoning tasks improves the systems' ability to think abstractly. The combination of generative AI and transformer architectures fundamentally improves CoT. Reasoning chains that are similar to human understanding become possible. In conjunction with chatbots, a dialogue-oriented AI opens up to carry out more complex interactions that require a deeper level of understanding and problem-solving skills.

Reminder of LLM training process in ChatGPT

Before showing how CoT is built into DeepSeek, it is worth recalling a generative AI such as OpenAI, which is built on the training of an LLM. As described already previously, the training process consists of three stages of training:

- *Pre-training*: In a first step, an LLM is trained with a huge amount of general-purpose knowledge. On this basis the model can predict the next letters, symbols or words (token) in a given sequence to get a reasonable context.
- *Supervised fine-tuning*: After pre-training, the model is fine-tuned on a dataset of instructions with several samples (SFT). A sample consists of pairs of an instruction and a response to

the instruction. As we already described in the section on machine learning, a neural network can be trained by these samples of instructions to become better and better with its answers.

- *Reinforcement learning*: Finally, the LLM is improved by feedbacks of reinforcement learning (RL). The model can be trained by feedbacks of humans (RLHF = Reinforcement Learning from Human Feedback) which needs gathering large-scale, high-quality human feedback. In the case of Reinforcement Learning from AI Feedback (RLAIF), the feedback of learning is realised by an AI model. The efficiency of RLAIF depends on the accuracy of feedback which needs high technical impact.

Reinforcement learning with DeepSeek-R1-Zero

The training of DeepSeek-R1-Zero starts with a pre-trained model DeepSeek-V3-Base with 671 billion parameters. Reasoning capacities can be improved through large-scale RL without SFT. To run reinforcement learning at a large scale, a rule-based reinforcement learning method is used without neural networks for SFT. Given a model to train and an input problem, the input must be given into the model. Corresponding to this input, a group of outputs is sampled. Each output consists of a reasoning process and an answer. A reinforcement learning method called Group Relative Policy Optimisation (GRPO) observes the sampled outputs (Fig. 2). It trains the model to generate the preferred options by calculating a reward for each output with predefined rules [44].

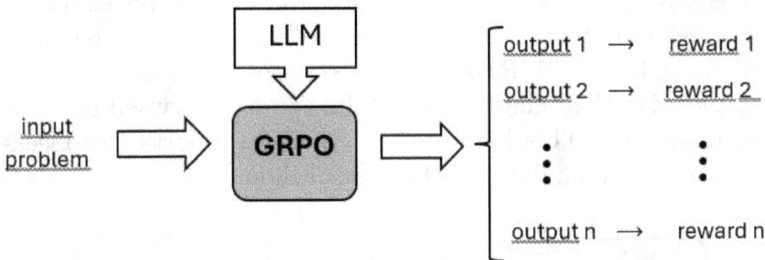

Fig. 2. For a given input, GRPO collects many outputs and instructs the model to prefer the best one by rewards for each output.

A first set of rules calculates a reward of accuracy. For problems in mathematics, that can be done with mathematical rules checking if the answer of the model is correct. In the case of a code model, a compiler can generate feedback based on the test cases.

A second set of rules generates rewards concerning the format. The model is instructed to respond in a certain format with a reasoning process within <think> tags and the answer within <answer> tags.

A rule-based procedure without using a neural network to generate rewards is more simple and reduces the costs of the training procedure, especially in the case of large models.

In DeepSeek-R1-Zero, we can already observe the 'Aha' moment during the reinforcement learning training. For example, in a mathematical question, the model starts its reasoning process. At a certain point, the model begins to reevaluate its solution. The model learns to reevaluate its initial approach. If necessary, it can even correct itself.

Training pipeline of DeepSeek-R1

DeepSeek-R1-Zero results often suffer from readability and inconsistencies in language formulations. In short, it must become more user friendly. Therefore, DeepSeek-R1 has been introduced with better performance. The training pipeline of this model consists of the following four phases:

- *Phase 1: cold start*

DeepSeek-R1 starts with the pre-trained model DeepSeek-V3-Base. The model is prepared by SFT on a small dataset of results collected from DeepSeek-R1-Zero. The dataset consists of thousands of samples. On this relatively small dataset a supervised fine-tuning phase is applied. This high-quality dataset supports DeepSeek-R1 to improve the readability in the initial model.

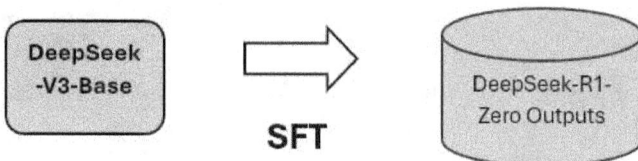

- *Phase 2: reasoning reinforcement learning*

In this phase, the same large-scale reinforcement learning of the previous model is applied to improve the reasoning capabilities of the model. The rewarding rules for the reinforcement learning process are defined by solutions of tasks such as in mathematical logical reasoning, coding and science.

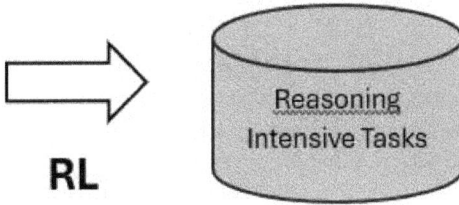

- *Phase 3: rejection Sampling and supervised fine-tuning*

In this phase, the controlling rules of phase 2 are used to generate many samples. With rejecting samples, only correct and readable samples are selected. The generative reward model DeepSeek-V3 is used to decide which samples should be kept. Training data of DeepSeek-V3 are also used in this phase. On this dataset, the model is trained by SFT. The questions of this dataset are extended across more domains.

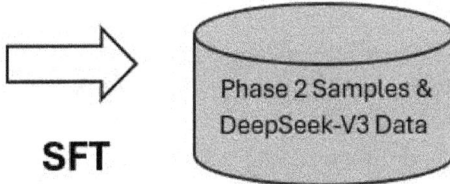

- *Phase 4: diverse reinforcement learning phase*

The final phase extends rule-based rewards to tasks such as mathematics, but other tasks can also be considered with an appropriate LLM.

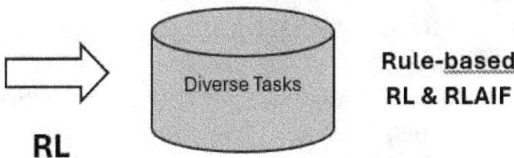

Finally, DeepSeek-R1 can be compared with OpenAI o1. Its efficiency is not only comparable but in some fields even surpasses the American competitor.

Applied technology of DeepSeek

The technology behind DeepSeek combines deep learning, natural language processing (NLP), big data integration and computer vision. These components work together to analyse complex datasets, recognise patterns and deliver results. Here, too, the close connection between complex systems and artificial intelligence, which is the subject of this book series, becomes clear.

- *Deep learning*: DeepSeek applies deep neural networks that are organised in many layers. Each layer specialises in analysing certain features or patterns in the data. As with deep learning in general, the first layer extracts basal features, while subsequent layers recognise increasingly complex relationships. In a medical application, one layer might process image data from a tumour analysis, while another layer examines text data from patient records.

The accuracy of the analyses is increased by convolutional neural networks (CNNs) for images and recurrent neural networks (RNNs) for time-dependent data. Transfer learning is used to adapt pre-trained models to specific problems. In this case, there are already many specialists for different areas in the system to whom the problem can be forwarded when solving the problem. This reduces computing time and increases efficiency.

- *Natural language processing*: The ability to understand natural language is a key feature of DeepSeek. Using NLP, the platform can analyse unstructured text data such as reports, scientific studies, emails or social media content. Key techniques include Named Entity Recognition (NER) to extract relevant terms such as names, places or drugs, as well as sentiment analysis to capture the mood or meaning of a text. DeepSeek also uses transformer models, such as those already used in ChatGPT.

These models make it possible to understand context in texts, which means that complex queries in natural language can be answered better. For multilingual data, the platform uses machine translation so that insights can be generated independently of the original language.

- *Big data integration*: DeepSeek can handle huge and heterogeneous datasets. Thanks to its cloud-based architecture, data sources from different areas can be integrated in real time. These include, for example, sensors in the Internet of Things (IoT), company databases or social networks. The use of distributed computing and clustering technologies ensures that the platform maintains its performance even with huge amounts of data. In addition, DeepSeek uses data pre-processing techniques to clean up unstructured or erroneous data and thus increase the accuracy of analyses.

- *Computer vision*: DeepSeek uses advanced computer vision algorithms to analyse image and video data. These include object recognition, pattern recognition and image segmentation, which enable the platform to extract detailed insights from visual information.

- *Augmented and virtual reality*: DeepSeek also uses augmented reality (AR) and virtual reality (VR) technologies to visualise data in immersive three-dimensional (3D) environments, which opens up new possibilities, particularly in research and development.

Invention and innovation of DeepSeek

DeepSeek's technological invention is the Large Language Model R1. As explained previously, this is an AI language model that is designed to understand, process and generate natural language. What makes it special is not only its performance but above all its resource efficiency: DeepSeek only invested around six million Euros in the development of R1, which is a fraction of the costs that companies such as OpenAI had to spend on comparable models such as GPT-4. In addition, R1 was trained with outdated Nvidia chips that were purchased before the US export ban. Despite these limitations, R1 achieves similar performance to the OpenAI and Google models and even outperforms them in some respects.

DeepSeek was founded in 2023 by Liang Wenfeng. Mathematically gifted, he studied engineering at Zhejiang University in Hangzhou. Instead of pursuing a university career, he and two other engineers initially earned money with algorithms that invested better than many small investors in China. Wenfeng was already intensively involved in machine learning for quantitative trading on financial markets during his studies. In 2016, he founded the hedge fund High-Flyer, which focuses on AI-supported trading strategies. This fund analyses financial market data in real time and automates decisions based on machine learning. Even back then, his motto was that only funds in which algorithms made investment decisions were genuine quantitative funds. High-Flyer was a milestone for Wenfeng, but his vision went further.

Liang Wenfeng has been deeply impressed by the career of the US-mathematician and hedge fund manager James Harris Simons (1938–2024) [45]. With his mathematical and engineering gifts, Liang Wenfeng also at first had gained in machine learning with hedge funds and modest capital. He founded the relatively small start-up DeepSeek to devote himself to the technical principles of AI from then on. In contrast to other (particularly American) start-ups, he was not interested in commercialisation, but in the technical fundamentals. The motto is neither to lose money nor to make huge profits. Instead, he criticised the prevailing mentality in the Chinese economy of making a profit and neglecting the real technical breakthroughs.

The Chinese start-ups copied open-source models from the American meta group. However, Wenfeng believes that this is only sensible if the goal is to develop applications with rapid development of profitable products. In his view, however, China should not be a free rider, but rather put itself at the forefront of innovation in the world. This includes basic research: 'Our goal is General Artificial Intelligence (General AI = GAI)!' This is essentially a philosophical vision. This shows how philosophically led basic research leads to long-term breakthroughs.

Nevertheless, he proved to be extremely successful commercially and shook up the AI market of the American tech giants. Even before DeepSeek was founded, he invested heavily in Nvidia graphics chips before the Unites States, under President Joe Biden, imposed an export ban on high-performance chips such as the

H800 and A800 models to China in October 2023. These chips are currently a technical prerequisite for training artificial intelligence systems. Despite these restrictions, DeepSeek has managed to develop an AI model with limited resources that can compete with the world's best technologies.

The troubling challenge for US policymakers and investors is that strict export controls have forced Chinese technology companies to become more self-reliant, leading to breakthroughs that might not otherwise have been achieved. This begs the question, why can a smaller start-up from China with a minimal budget compete with billion-dollar projects from the United States? If there is a cheap AI model, this calls into question the profits of competitors who have already invested billions in a more expensive AI infrastructure. In the end, it was human intelligence and innovation that led to new ways of training AI more efficiently and cheaply under the pressure of restrictions.

Application fields of DeepSeek

DeepSeek's versatility and performance make the platform a tool in a variety of industries where huge amounts of structured and unstructured data need to be processed.

Example: Medicine and healthcare
In the healthcare sector, DeepSeek can accelerate and refine diagnoses and develop new therapies. Examples include early detection of diseases, personalised medicine, drug development and pandemic monitoring.

Example: Climate research
In climate research, complex patterns in environmental data can be recognised and precise prediction models created. Examples include analysing weather data, early detection of natural disasters, sustainability strategies and biodiversity protection.

Example: Financial sector
In the financial sector, risks can be minimised, market trends predicted and investments optimised. Examples include market analyses and forecasts, risk management and fraud detection.

Example: Manufacturing industry

In the manufacturing industry, DeepSeek helps companies to make their processes more efficient and sustainable through predictive maintenance, quality control, supply chain optimisation and energy efficiency.

Future of DeepSeek

The potential for further development and innovation is far from exhausted. In the coming years, DeepSeek could play a role in how data is utilised and complex problems are solved. A great future prospect for DeepSeek would be the integration of quantum computing. While conventional computers are reaching their limits, especially when analysing extremely complex datasets, quantum computers could overcome these limitations. By utilising quantum computing technologies, DeepSeek could become even more powerful and solve problems that are currently considered unsolvable.

While AI technologies are still limited in many cases to technologically advanced countries, DeepSeek could contribute to the spread of AI as an open-source model. With its cloud-based architecture and flexible pricing models, the platform could also become accessible to developing countries or smaller companies. This would not only promote global innovation but also help to reduce social inequalities.

Quantum AI systems

The quantum world has long since arrived in everyday life without many people realising it. This includes transistors, diodes and lasers, which have become an integral part of everyday devices and are used as a matter of course in measurement and communication technology, industrial production, medical technology, consumer electronics and 3D printing. What the quantum technologies of this first generation have in common is that quantum effects are only utilised indirectly.

We are currently living in the second generation of quantum technology, in which the basic principles of quantum mechanics are specifically implemented in quantum mechanical devices. These include the first prototypes of quantum computers, classic supercomputers with quantum simulation, quantum cryptography and quantum communication, quantum sensors and quantum measurement technology. To solve a specific task, a special quantum computer from Google was able to prove the superiority of a quantum computer over a conventional supercomputer for the first time in 2019 [46]. In the same year, IBM Q, a quantum computer based at least in principle on the architecture of a general-purpose computer, had already been presented [47].

Nevertheless, the quantum world is still often perceived as mysterious by the public. What Einstein saw as a spooky effect in 1935 has long since become the basis for revolutionary quantum communication in fibre optic networks and satellite technology, heralding a future quantum Internet. Quantum computers as multipurpose computing devices are just the tip of the iceberg of a technology that is gradually spreading as the network of our civilisation. That is why it is misleading to speak of 'disruptive' technology that suddenly appears. It only seems that way to those for whom this technology is incomprehensible.

Fundamentals of the quantum world

This makes it all the more urgent to understand the fundamentals of the quantum world as the background to this technology [48]. Understanding the basics and interrelationships also requires an understanding of mathematical language. It is ultimately an expression of 'common sense', which creates conceptually precise instruments to calculate experiments and enable technology. Calculation methods are converted into algorithms and computer programs.

The quantum world with its mathematics enables algorithms and procedures that are incomparably richer, faster and more effective than classical computers.

A quantum bit contains not only two alternative information states of bits 0 and 1 but all states between 0 and 1. In technology, such 'superpositions' of states lead to quantum parallelism, which enables simultaneous calculation of all states and thus considerable acceleration compared to successive calculation steps in classical computers.

> Quantum mechanical 'entanglement' of distant locations enables instantaneous quantum communication. In the quantum world, distributed probabilities enable 'tunnelling through' obstacles, avoid classical detours and thus significantly accelerate computing processes.

From the quantum world to quantum computing

Quantum computing brings together physics, computer science, logic and mathematics. Whether the world itself is a computer (Leibniz) or can at least be simulated by a computer (Feynman), this vision runs like a red thread through the history of ideas in modern times.

None of this is witchcraft but results step by step from the fundamentals of the quantum world. As Richard Feynman, the pioneer of the quantum computer, correctly stated, it is the actual and comprehensive physical reality. Classical physics only describes an approximate section of the macroscopic world. In this sense, quantum physics is a natural approach to the world and not a question of magic tricks like 'Schrödinger's cat' with its whimsical lives [49]. In addition, thinking in terms of statistics and probabilities is very much in line with everyday experience, which is by no means determined like classical mechanics.

Dangers and security through quantum computing

Quantum technology and quantum computers will increasingly become part of everyday life. However, quantum technology, quantum computing and artificial intelligence are not only opening up new possibilities but also harbour new dangers. In addition to absolutely secure quantum cryptography, quantum communication could also be used for military purposes, for example, for global drone control. Quantum computers, which beat classic supercomputers, enable far more and more effective control and manipulation worldwide. The call for early technology design, which I already made in my AI book, therefore becomes even more urgent. However, this also makes it clear that the demand for sustainable technology is indispensable in the face of global crises. Ultimately, we need a future policy in which the central role of sustainable science and technology in the 21st century is understood and made the basis for political judgement and decision making.

Generations of quantum technology

The quantum computer is just the tip of the spear of a broad spectrum of quantum technologies that are revolutionising the economy and society. Following basic research into quantum mechanics at the beginning of the 20th century, technical components based on the effects of quantum physics have been developed since the end of the 20th century. These include transistors, diodes and lasers, which can be summarised as first-generation quantum technologies and have become an integral part of everyday devices. The laser was also initially a basic quantum physics problem that is now used as a matter of course in measurement and communication technology, industrial production, medical technology and consumer electronics, right through to 3D printing. What first-generation quantum technologies have in common is that quantum effects are only utilised indirectly.

We are currently living in the second generation of quantum technology, in which the basic principles of quantum mechanics are specifically implemented in quantum mechanical devices (Fig. 3). These include the first prototypes of quantum computers, classical supercomputers with quantum simulation, quantum cryptography and quantum communication, quantum sensor technology and quantum measurement technology (metrology).

To solve a specific task, a special quantum computer such as Sycamore with 54 qubits was able to demonstrate 'supremacy' for the first time in 2019, i.e. the superiority of a quantum computer over a classical supercomputer. In the same year, IBM Q, a quantum computer with 20 qubits, had already been presented, which at least in principle is based on the architecture of a universal multi-purpose computer. Although both are not yet directly usable for commercial applications, they were built by two of the major IT companies such as IBM and Google as milestones on the way to a universal quantum computer with supremacy.

From the universal digital computer to the universal quantum computer

In the age of digitalisation, the first classic digital universal computers were available in 1941 with Zuse's Z3 and in 1945–1946 with John von Neumann's ENIAC. Both opened up the development of a broad spectrum of digital technology. The third generation of quantum technology would begin with a universal quantum computer (Fig. 3).

Such a universal quantum computer, which could, for example, implement the Shor algorithm for factoring large numbers, would require millions of qubits. Due to the sensitivity of such quantum algorithms to noise, it would have to be able to perform highly complex error corrections. It has already been shown that a quantum computer with over 50 qubits can solve tasks at a speed that is not feasible for the fastest classical supercomputers. This means that there are already quantum computers that can perform tasks that classical computers are not capable of. On the other hand, however, they are not yet large enough to realise a fault-tolerant application of the known quantum algorithms.

The American computer scientist John Preskill called this era between quantum computers with 50–100 qubits (e.g. Google's Sycamore) and the first universal quantum computer with 1,000,000 qubits and more 'Noisy Intermediate-Scale Quantum' (NISQ) [50]. It is 'noisy' because there are not enough qubits available for error correction, and 'intermediate-scale' because the number of qubits is sufficient for proof of supremacy but not yet sufficient for a

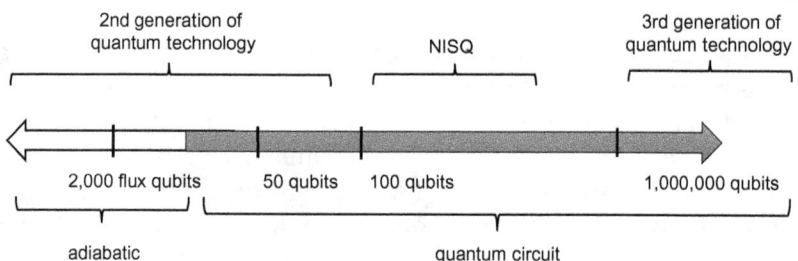

Fig. 3.   Phase transitions of quantum technology.

universal quantum computer. The current NISQ era thus describes the transition from the second to the third generation of quantum technology (Fig. 3).

A look at the emergence of the first digital computers shows the enormous challenges that still need to be overcome before the first universal quantum computers with supremacy can be developed in the third generation of quantum technology. First, there is the hardware technology, in which superconducting materials are emerging as a key technology. The associated low-temperature physics seems most likely to be able to guarantee superpositions and entanglements with sufficient coherence time.

A less common approach is ion traps with ionised atoms in a vacuum. It is true that the qubits are more durable than in superconducting circuits. However, the reaction times for control processes are slower. In photonics, entangled photons from single photon sources are used. The major advantage is that they function at room temperature. In contrast to superconducting materials, this approach is less well developed to date. Semiconductors or quantum dots allow the expansion to larger devices (scalability). They also continue the already highly developed semiconductor technology. However, like superconductors, they require very low temperatures. Also worth mentioning are what are known as topological quantum computers, which use exotic elementary particles as computing units. Theoretically, they are said to be less prone to errors, but this approach is the least developed so far.

Adiabatic quantum computer in commercial application

In the second generation of quantum technology, quantum computers that are not based on universal quantum circuits modelled on classic digital computers are already in commercial use. This refers to adiabatic computers or quantum annealers, which are used commercially by companies such as D-Wave (Fig. 3) [51]. In contrast to logical circuit models, adiabatic computing is orientated towards phase transitions of energy states, as known from thermodynamics. The solution of an optimisation problem is linked to the global minimum of an energy function, which is to be achieved by slow (adiabatic) cooling of the energy system.

One example is the annealing process in the steel industry, in which phase transitions are achieved by heating, holding and cooling metals. The corresponding calculation processes are therefore also referred to as simulated annealing. In the quantum version, adiabatic computing utilises the quantum states of elementary particles to store computing operations and data. Quantum annealing refers to quantum algorithms for solving optimisation tasks in which quantum fluctuations are used to support the search for the global minimum. Its applications include faster calculations of traffic flow and the solution of other optimisation tasks.

Phase transitions of quantum technologies

As in classical computers, the various levels of software are based on the hardware. On the theoretical side, mathematical algorithm theory based on the formalism of quantum mechanics (e.g. algorithms by Shor, Grover) is the most advanced [52]. It must be related to models of quantum circuits, which in turn are based on the physical technology of the hardware. However, the various programming levels play a major role right up to the user. There are initial approaches for commercially and scientifically usable programming environments, as the examples of Cirq from Google and Qiskit from IBM show [53].

Quantum simulations are already highly developed in the second generation. They facilitate, for example, the detection of material defects that have electromagnetic causes. In general, optical material properties can be determined in this way. In order to produce drugs efficiently and cost-effectively, complex molecular structures must be calculated. Quantum simulation will prove to be indispensable for the increasingly complex challenges of biotechnology. Finally, quantum computers and quantum simulation will make artificial intelligence applications more efficient.

In the age of big data, conventional computers are clearly reaching their limits in the application of learning and search algorithms.

In the development of digitalisation, first came the universal digital computer in the 1940s and 1950s and then digital communication technology up to the digital Internet in the 1990s. In the development of quantum technologies, it is becoming apparent that quantum communication with fibre optic cables, satellite technology and/or superconductors can be used commercially at the same time or even earlier than the universal quantum computer. This is due to the already highly developed technology of fibre optic cables, satellites and superconductors in solid-state physics. However, quantum communication with fibre optic networks requires 'amplifiers' (quantum repeaters) for long distances of thousands of kilometres, which are still outstanding as technical standards [54]. Whether at least a proof-of-concept will be technically available in 1–2 years' time remains to be seen. In any case, quantum repeaters will be indispensable building blocks for the commercialisation of quantum communication with fibre optic networks. In contrast, satellite technology may be expensive, but it is already technically mature.

Innovation dynamics of quantum technology

Another key challenge is securing data communication in good time using quantum cryptography before quantum computers crack the classic codes. Quantum mechanical encryption guarantees absolute security for data transfer — an indispensable prerequisite for financial transactions, for example, especially in the future use of digital currencies. Manipulation of personal data, such as patient records in social networks, also requires better security, which quantum cryptography can provide. In the age of digitalisation, civilisation has also become increasingly dependent on digital infrastructures. Cyberattacks can disrupt the energy supply or logistical supply chains at any time, for example. In this case, too, quantum communication and quantum cryptography will be indispensable.

Highly developed quantum technologies with commercial applications already include quantum sensor technology and quantum metrology. Quantum mechanical devices can be used to measure physical quantities such as time, position, speed, pressure and temperature, magnetic and electric fields or gravity with extreme precision. Atomic clocks based on atomic quantum states have been in use for decades as a time reference.

Examples include navigation systems such as Galileo in Europe or GPS. Extremely precise clock measurement in the quantum range is necessary to improve time scales and satellite navigation or to synchronise large data networks and radio telescopes. Safe and more accurate navigation systems for autonomous driving or controlling global logistics systems, for example, will have to replace traditional GPS. In medical technology, quantum sensors measure more accurately and with higher resolution. This means that the smallest tissue changes can be recognised at an early stage, for example, in cancer diagnostics, as can neuronal processes in the brain. In robotics and artificial intelligence, quantum sensors are taking the place of conventional sensors and increasing the potential many times over.

Measurements of gravitational forces, magnetic and electric fields with quantum sensors will be indispensable for Earth monitoring. Ultra-precise gravimeters based on the interferometry of cold atoms are being used. This quantum technology can also be used to locate mineral resources. Finally, precise measurements are used to support early warning systems in climate protection. Quantum sensors, which can also be used at room temperature and are also becoming smaller and cheaper, are the goal of market-orientated quantum technology.

To summarise, the development of the quantum computer also shows that in research, the journey can be the destination: it is not only the ultimate goal of the universal quantum computer that is revolutionary, a broad spectrum of new innovations is being developed along the way. For example, quantum-based measurement technology has great economic potential, ranging from navigation, geology and Earth observation to medical diagnostics, industrial precision measurement technology and military technology.

From basic research to technology design

The example of quantum technology clearly shows how the success of a technology's application can be deeply rooted in the philosophical foundations of science. It began with fundamental discussions and thought experiments (e.g. entanglement), which were eventually translated into laboratory experiments. Ultimately, it is about the transition from experimental set-ups in the laboratory to robust, reliable and cost-effective devices. This requires supporting industrial engineering technology that makes the construction of these devices possible in the first place. This is referred to as 'enabling engineering' [55].

Typical examples are quantum-compatible data acquisition, fast electronics for data processing with high time resolution, low dead times, data throughput-optimised, parallelised, etc. with associated software. The first generation of quantum technology includes lasers and detectors. Enabling technology also includes materials, components and quantum technology devices and processes for positioning and implanting individual atoms, ions or molecules, vacuum technology and optical precision construction.

A mature enabling technology influences the future of an innovation. The branch of development that can build on already highly developed technology with standards and norms often prevails. In the end, markets create a push and pull effect to steer technical developments in certain directions. Markets are no longer just about demonstrating feasibility in principle (e.g. proof of supremacy from Google's Sycamore), but about the turnover and sale of commercial goods.

The development of the quantum computer is embedded in the development of quantum technologies, which in turn are part of the global trend towards digitalisation. The concept of the quantum computer is therefore just a beacon of research development. However, the promise of the future 'quantum computer' does not just mean a single device like Zuse's Z3 and von Neumann's ENIAC, which will be available at a certain point in time, but the broad avenue of technological development that is already changing markets and civilisation.

Phase transitions instead of 'disruption'

The sociological term 'disruptive' technologies is therefore misleading because it suggests a sudden event that only changes the world when it occurs. In fact, it is a broadly diversified development that appears to be continuous if you take a closer look at the facts. Only those who do not know or do not understand the basics and backgrounds must find the development abrupt. In-depth philosophical analysis and reflection are therefore required in order to be able to strategically assess and evaluate such technology trends correctly.

Why philosophy? Since ancient times, it has been the origin of the sciences, which have become increasingly specialised over the centuries [56]. Even Newton, the founder of modern physics, had a chair in natural philosophy, while his compatriot Adam Smith, the founder of modern economics, had a chair in moral philosophy. Newton called his main work *Principia Mathematica Philosophia Naturalis*, i.e. mathematical principles or foundations of natural philosophy. Since the Aristotelian tradition, 'natural philosophy' has encompassed what is now called natural science. Observation and logical analysis, even the first mathematical models, existed before Galileo.

Since Galileo's time, however, measurement technology, observation and experimentation have been systematically combined with mathematical models. However, this was no more 'disruptive' than the heliocentric model of Copernicus. Apart from the fact that Greek astronomers were already discussing the possibility of a heliocentric planetary model, the development of natural science was extremely diversified over a long period of time: the basic concepts and methods of natural science did not suddenly appear and were proclaimed like the Ten Commandments by a Moses of science.

Thomas Kuhn's distinction between 'scientific revolutions' and 'normal science' is, at best, a simplification and justified by a sociologising and psychologising perception. The criticism of superficial talk of 'disruption' also applies to the emergence of quantum physics in the 20th century. Historically, quantum mechanics had to develop from familiar concepts, procedures and ideas of classical physics, in which generations of physicists were educated until the beginning of the 20th century. However, classical mechanics by no means corresponds to our everyday ideas, as is always claimed. That bodies reduced to points should move in a vacuum on ideal mathematical curves is a highly abstract model, which even today must seem completely abstract and detached from everyday experience to people before their first physics lessons. In fact, Aristotelian physics describes our everyday experience in phenomenological terms, according to which falling solid bodies sink to the ground in the air around us in a highly complex way, as if in a liquid. The model of classical mechanics was only easy to calculate once the mathematical calculations and algorithms were available, in which every A-level student must first be trained.

However, anyone studying physics starts with quantum mechanics in the first semester, learns the linear algebra and functional analysis of their mathematical calculations and solves lots of exercises. This practice creates familiarity and habits with the underlying physical models. Classical mechanics now appears as a simplified model that only has approximate value in special cases. The quantum world is taken for granted as the actual reality.

Perspective of the philosophy of science and technology

This is where a modern philosophy of physics and technology comes in, with which the understanding of the concept of states in the quantum world, the meaning of superposition, entanglement and tunnelling can be explained and deepened. There is nothing 'unnatural', 'puzzling' or 'disruptive' about it. Even in classical mechanics, we are familiar with the phenomenon of only doing the maths without understanding the conceptual relationships. This becomes apparent at the latest when problems arise that deviate from the usual routine.

In quantum technology, too, it is dangerous to lose sight of the understanding of the basics. In the end, we end up continuing with previously successful routines and become 'operationally blind' to new and unusual technical paths that open up the potential of the basic concepts of quantum mechanics. Philosophy aims to understand the fundamentals and thus contributes to a better understanding of the application possibilities. Philosophy of technology, which deals with the goals of technology design, builds on this.

Today, as in the times of Aristotle, philosophy still enquires into the principles and foundations of our knowledge and its interdisciplinary connections in the various disciplines in order to be able to make decisions and act responsibly. Since ancient times, logic, the foundations of science and ethics have belonged together in philosophy. Problem- and practice-oriented networking with the sciences is the special profile of philosophy in the globalised knowledge society.

It is crucial that philosophy and philosophy of science are anchored in the individual subjects of engineering, natural sciences, social sciences and economics. Only constant research and teaching contact can prevent philosophers from taking off into the clouds of abstraction, burying themselves in the history of the discipline and losing contact with science. This is the only way to stimulate the necessary fundamental discussion in the sciences on the part of philosophy. However, this requires appropriately trained philosophers in, for example, mathematics, computer science, physics, biology, sociology and economics, who are accepted as competent in these disciplines.

Philosophy should therefore by no means be reduced to its minor subject of ethics, as is unfortunately often the fashion today. Aristotle wrote his little booklet of *Nicomachean Ethics* on this subject, as well as extensive compendia on the fundamentals of physics and logic of his time. Only those who have understood the fundamentals of science can venture into innovative new territory in technology and, building on this basic knowledge, speak competently about ethical issues [57]. Ethical challenges of technology can therefore only be met if the fundamentals are understood.

However, as quantum technology advances, there is a growing risk that quantum computers will crack the security codes on which financial, economic and defence systems are based. Machine learning verification methods must therefore be extended to the possibilities of quantum computers and quantum communication. On the other hand, the fundamentals of quantum mechanics also open up absolutely secure quantum cryptology for quantum communication networks. Ultimately, the goal must be standardisation (ISO standards internationally) of quantum technology and quantum computers, as is already the case with conventional digitalisation technology and is being sought for artificial intelligence tools.

Market potential of quantum computers and quantum technology

In addition to the economy, the state and civil society also have a major responsibility. The transition from basic research to industrial application often only succeeds in research alliances and with the support of research portals, which must also be flanked at the state level [58]. It should be noted that the maturity levels of the various quantum technology fields differ.

What the second-generation systems have in common is that they are only just beginning to be commercialised. As a result, there are no fully developed value chains yet. Companies and research groups are currently analysing the implementation of laboratories in practical applications. Based on previous trends in research and technology, it is possible to estimate market potential and value chains for the future. For quantum computers alone, a market worth hundreds of billions is predicted in the medium and long term. As already explained, the various side developments and the associated market potential on the way to the quantum computer are at least as interesting as the quantum computer itself.

With its theoretical market potential, the quantum computer would leave all second-generation quantum technologies behind. In an optimistic scenario study, it is assumed that a market will emerge from 2030 that will grow to USD 57 billion in 2035 and USD 295 billion in 2050. In the more conservative case, the large market shares are not assumed until a later date (2040).

However, even then, strong growth is forecast from USD 6 billion to USD 263 billion within 10 years. Another study estimates the short- to medium-term market development from USD 0.5 billion in 2023 to USD 5 to 10 billion in the period from 2020 to 2025 to USD 23 billion in 2025. More conservative forecasts compare the long-term market potential of hardware for quantum computing with the current size of the market for supercomputers (USD 50 billion). For the 2020s, the market for NISQ quantum computers (Fig. 3) is already estimated to be worth several hundred million USD [59].

Cost savings and sales growth through quantum computers are estimated at USD 2–5 billion over the next 3–5 years. Over the next 10 years, it is expected to be worth USD 25–50 billion. After the end of the NISQ era in more than 20 years, a market potential of USD 450–850 billion is assumed. In particular, the integration with artificial intelligence, which is expected to take place during this period, could lead to sales that cannot yet be estimated in monetary terms. The business model is not expected to involve the mass production of quantum computers in the near future. Instead, IBM's 'quantum computing-as-a-service' business model will prevail, according to which customers can use the programming environment of a quantum computer.

Quantum computing tasks are often embedded in classical computing. It is therefore foreseeable that quantum computing will be offered as a subroutine of conventional data centres with supercomputers. Commercially, there are already companies such as D-Wave that sell applications with adiabatic computers.

Social acceptance of quantum technology in technical communication

As might be expected, the greatest acceptance of quantum technology is in medical applications such as medical technology and diagnostics. The complexity of the human organism will not be manageable without the computing power of quantum computers. In order to decipher the codes of viruses, for example, bioinformatics requires novel computational models and immense computing power. This is the only way to reduce the time-consuming and lengthy testing of drugs and vaccines in order to reach the goal more quickly. In the case of a universal quantum computer, quantum algorithms that can simulate evolutionary processes are conceivable. Quantum computers and quantum technology would be invaluable as early warning systems for potential viruses or tumours. The medical role of quantum sensors can also be emphasised in this context.

Quantum technologies will open up new avenues in materials research that will also directly affect consumers. The development of electrification and digitalisation, which provided many everyday objects, can be recalled here. Superconducting materials and semiconductor technology will become an integral part of civilisation's infrastructure. These connections between wide-ranging research developments must be communicated to the public [60]. The question 'When will the quantum supercomputer arrive?' becomes secondary in comparison.

Of course, such acceptance studies cannot assume that the respondents have understood what quantum technology actually means. However, the statement that everything will work much faster, more accurately and more securely is seen as positive, especially in medicine. In the UK, however, the population fears that expensive high-tech will make medical services more expensive, while military applications are not a cause for concern [61]. In Germany, on the other hand, there has been a critical reaction to military applications. There is also often uncertainty about what absolutely secure quantum communication means. If quantum computing leads to an increase in the possibilities of machine learning, fascination and fear will be transferred from the topic of artificial intelligence to quantum computing.

Public fears centred on the issue of communication security should be taken very seriously. Factual information about the basics of quantum communication is not the only thing that can help here. On the one hand, the benefits of communication security must be emphasised. On the other hand, the risks of the emerging surveillance possibilities must be clearly identified as well as ways in which political participation can be strengthened. However, risks also relate to the population's growing need for security. For example, quantum communication can be used for faster and more secure communication during pandemics and other global disasters.

Possible economic and therefore professional changes combined with managing the fears of increasing surveillance and manipulation of data are crucial for acceptance among the German population. However, quantum technologies are not expected to have a drastic impact on the labour market. This discussion is familiar in connection with automation and the use of robots in industry. Quantum technologies do not directly threaten jobs.

On the contrary, it is assumed that new highly skilled jobs will be created. The public perception is that quantum computing will only increase the good and problematic aspects of digitalisation on a gigantic scale. Similar to artificial intelligence, the technology is perceived as mysterious and dark. Some still feel that quantum physics reminds them of magic tricks.

The only thing that can help here is education, which must begin at school. Quantum technology is preparing to penetrate all areas of civilisation, just as electricity did in its day. Quantum computers will be part of everyday life. At this point, the basic understanding that was called for in connection with the philosophy of science and technology must be combined with technology communication and education. Quantum physics must no longer appear bizarre, mysterious and counterintuitive to the public. The analogy with the frequently used magic tricks is only apt insofar as behind every magic trick there is also an understandable explanation.

Quantum physics thinking is the norm in nature. Its approach to statistics corresponds to our everyday experience. This background also provides an understandable approach to basic concepts such as superposition and entanglement. Even the mysterious 'tunnelling through' becomes comprehensible with a thorough understanding of a probability distribution. School and education must clearly convey a basic understanding of this way of thinking and also train the fundamental techniques, just as was done for centuries with classical mechanics.

Neuromorphic AI systems

Classical AI research is based on the capabilities of a program-controlled computer which, according to Church's thesis, is in principle equivalent to a Turing machine. According to Moore's law, gigantic computing and storage capacities have thus been achieved, which only made possible the AI services of the WATSON supercomputer. But, the power of supercomputers has a price that the energy of a small town can match. All the more impressive are the human brains that realise the power of WATSON (e.g. speak and understand a natural language) with the energy consumption of a small lamp. By then, at the latest, one is impressed by the efficiency of neuromorphic systems that have evolved in time. Is there a common principle underlying these evolutionary systems that we can use in AI?

Biomolecules, cells, organs, organisms and populations are highly complex dynamic systems in which many elements interact. Complexity research in physics, chemistry, biology and ecology deals with the question of how the interactions of many elements of a complex dynamic system (e.g. atoms in materials, biomolecules in cells, cells in organisms, organisms in populations) can lead to orders and structures as well as to chaos and decay.

From complex dynamical systems to evolution

Generally, in dynamic systems, the temporal change of their states is described by equations. The state of motion of a single celestial body can still be precisely calculated and predicted according to the laws of classical physics. For millions and billions of molecules, on which the state of a cell depends, it is necessary to resort to high-performance computers that provide approximations in simulation models. It is remarkable that complex dynamic systems obey the same or similar mathematical laws in physics, chemistry, biology and ecology.

The basic idea of complex dynamic systems is always the same: only the complex interactions of many elements generate new properties of the overall system that cannot be traced back to individual elements. Thus, a single water molecule is not 'moist', but a liquid due to the interactions of many such elements. Individual molecules do not 'live', but a cell does because of their molecular interactions. In systems biology, the complex chemical reactions of many individual molecules enable the metabolic functions and regulatory tasks of entire protein systems and cells in the human body. In complex dynamic systems, we therefore distinguish between the micro-level of the individual elements and the macro-level of their system properties. This emergence or self-organisation of new system properties becomes computable in systems biology and can be simulated in computer models. In this sense, systems biology is a key to the complexity of life.

From complex dynamical systems to network physiology

The human organism consists of a complex hierarchy of different physiologic and organ systems. Each of these systems has its own structural organisation and functional complexity. They are governed by non-linear dynamics of phase transitions leading to different states. In medicine, health and disease are causally based on these complex physiologic states. But, in general, modern medicine is professionally highly specialised. Therefore, physicians often follow a reductionistic program and reduce health and disease to the special organisation and dynamics of individual organ systems which they are specialised in.

However, the human organism is an integrated complex network of multi-component physiological systems. Each not only has its own structure and regulations, but interacts continuously with others to coordinate its functions. It is this network of coordinated interactions among organs that generates the physiological states to maintain the health of the whole organism. Physiological interactions can be observed at different levels of integration and across spatiotemporal scales of organ functions. In the case of health, the physiological network optimises and synchronises its dynamics at the level of the whole organism. Mathematically, the various signalling pathways of interaction can be modelled by stochastic and non-linear feedbacks.

Physiologic states at the level of organisms are, for example, wake and sleep, stress and anxiety, cognition, and consciousness and unconsciousness. They are the result of coordinated network interactions among systems and subsystems. Local perturbations in this complex network of signal communication can initiate a cascade of failures leading to a collapse of the entire organism, such as in the case of sepsis, coma and multiple organ failures. They are states of systemic diseases without a single cause. The principles of complex systems dynamics are necessary to explain how diverse systems and subsystems in the human body dynamically interact as a network and integrate their functions to generate physiological states in health and disease.

The new multi-disciplinary field of network physiology integrates the different research fields and focuses on the coordination and network interactions among diverse organ systems and subsystems with the principles of non-linear systems dynamics. The following survey on the interdisciplinary field of network physiology as complex systems science is based on the survey article [62]. A fundamental problem in physical, biological and physiological systems is to understand global states which emerge out of networked interactions among dynamically changing local states of subsystems. In early modelling approaches of the cardiovascular system, for example, electrical circuits were used to simply sum up individual measurements from separate physiologic experiments. But they could not explain the non-linear emergence of physiologic behaviour on the macro-level of medical treatment. These complex states arise from interactions of multi-component cellular and neuronal subsystems that regulate each organ in the human body with scale-invariant and non-linear output signals. This dynamics is further connected by various coupling and feedback interactions between organ systems that continuously vary in time.

The framework of computational neuroscience and systems physiology, which focuses on neuron-to-neuron signalling and on integration within organ systems, is not sufficient. The challenge is bridging from micro-level subcellular and cellular insights on genomic, proteomic and metabolic interactions to macroscopic epidemiological observations. There is a wide gap in research efforts and knowledge at the mesoscopic level of horizontal network interactions across organ systems and subsystems essential to maintain health. The new field of network physiology has emerged to fill this gap. The question is how physiological systems synchronise and integrate their dynamics as a network to optimise functions and to maintain health.

From the practical point of view in medicine, the complexity of network physiology is challenging. The complex, multi-scale dynamics of organ systems make it extremely difficult to identify and quantify the network of organ interactions. Collecting such data in both ambulatory and clinical environments is particularly problematic because medical devices are often not interoperable.

A key question is how physiological states and functions emerge out of the collective network dynamics of integrated systems. While network structure may play a role in generating various states and functions, different global behaviours at the organism level can emerge from the same network topology due to changes of network nodes and modulations in the network links. In a generalised methodology of complex network dynamics, nodes represent subsystems with diverse dynamics, interacting through different forms of coupling that continuously change in time with transitions across states and conditions. Novel concepts and approaches derived from recent advances in network theory, coupled dynamical systems, statistical and computational physics, biomedical informatics, signal processing and biological engineering provide new insights into the complexity of physiological structure and function in health and disease, bridging across levels of integrating subcellular signalling with intercellular interactions and communications among integrated organ systems and subsystems. These advances are first steps for building up a new research field of network physiology.

In traditional complex network theory, edges and links are constant and represent static graphs of association. Network physiology has to consider the complex dynamics of individual systems as network nodes. Links represent organ communications in real time and the evolution of organ interactions with time. Collective network behaviour emerges from changes in physiologic states and conditions. This new research field integrates empirical and theoretical knowledge across disciplines with the aim to understand how diverse organs, physiological systems and subsystems dynamically interact as a network from the cellular to the organism level to produce various physiological states and functions in health and disease.

In classical graph theory, nodes and links are static and represent statistical correlations and dependence rather than dynamical coupling. Dynamical aspects in classical network theory arise from diffusion processes of flow on a fixed network. A fixed network topology defines networks' function to transmit information. In contrast, in network physiology, links represent dynamical coupling and coordination between diverse systems and subsystems. A fundamental question is how to quantify, predict and control emergent global behaviours in temporal multiplex networks of diverse dynamic systems interacting simultaneously through various functional forms of coupling. In such adaptive networks, different global behaviours can emerge from the same network topology due to minor temporal changes in the dynamics of a node or in the functional form of a link.

Nodes are not identical but represent diverse dynamical systems with diverse forms of coupling which continuously change in time. Such investigations are not simply an application of established concepts and approaches in complex networks theory to existing fields of biomedical research. The new field of network physiology shifts the focus from single organ systems to the network of physiologic interactions with the aim to uncover basic laws of communication and principles of integration in networks of diverse physiological systems and their role in generating global behaviours at the organism level.

Under the umbrella of complex systems science, network physiology methods of statistical physics, applied mathematics, informatics and network theory are used to solve problems in systems biology, neuroscience, physiology and medicine. New computational and analytical approaches are needed to extract information from complex data, to infer transient interactions between dynamically changing systems and to quantify global behaviour at the organism level generated by networks of interactions that are functions of time.

In network physiology, empirical and theoretical, basic and clinical research are integrated. Examples of empirical and theoretical research are advanced methods for non-linear dynamics and synchronisation, theory of dynamical systems and adaptive networks, data-driven models of complex systems and their interactions, control theory in dynamic networks, information theory for coupling inference and causality for non-stationary and non-linear systems, new generation of data-intensive AI and machine learning algorithms for inference of network dynamics and function, biomedical engineering of sensor networks and human–machine interfaces. Examples of basic and clinical research are basic physiology and clinical medicine, networks of cell assembles, networks of the autonomic and peripheral nervous systems, brain structural and functional networks, biomechanical networks in tissues, networks in the cardio-vascular and respiratory systems, networks of skeletal muscle groups and muscle fibres, pairwise and network interactions of organ systems and subsystems. For practical physicians, these complex phenomena manifest in ageing, exercise and sports, as well as in numerous clinical and pathological conditions of multiple physiological systems in the human body, such as traumatic brain injury, cardiac arrest, neurodegenerative disorders, diabetes, sepsis, coma and multiple organ failure.

The field of network physiology also involves bioengineering and platforms for synchronised high-frequency recordings from different physiological systems in clinical environment and networks of wearable sensors for continuous measurement of physiological parameters. Integrated networks of clinical monitoring devices and wearable sensors are crucial to explain causal pathways of dynamical interactions in networks of physiological systems and to predict critical events. The human body generates continuous streams of physiological signals as output dynamics of various systems and physiological parameters. Machine learning with AI algorithms are necessary to identify the topological structure and the temporal dynamics of physiological networks. The target is to predict hierarchical reorganisation and cascades of breakdown in dynamic networks and to classify states, functions and conditions based on network physiology maps.

Physiological systems operate on time scales from milliseconds to hours and provide different types of output patterns from oscillatory, periodic and quasi-periodic up to chaotic behaviour. Each integrated physiological system exhibits multiple simultaneous interactions and different forms of coupling with other systems, where interactions among systems vary in time. This leads to a transient multi-layer network structure consisting of distinct physiologic networks. Therefore, global network dynamics of the entire organism cannot be simply expressed as a sum of the behaviours of individual systems. Local changes can initiate cascades of different reactions leading to global changes, even when network topology remains unchanged.

Main research tasks of network physiology concern the nonlinear dynamics of brain–organ and organ–organ networks to understand their physiologic control. How can robust associations of network structure and dynamics be connected with physiologic states and functions at the organism level? Basic universal principles of integration in networks of diverse physiological systems are necessary to explain interactions between motifs, modules, subnetworks and networks formed by physiological systems at different levels and time scales. Physiological states and functions emerge from network interactions among diverse systems. Different degrees of coupling, link intensity distribution and coordination between systems dynamics are necessary to realise a physiological state at the organism level. Physiological networks must hierarchically reorganise with transitions from one physiological state to another in response to changes in autonomic regulation.

Rigorous mathematical and algorithmic techniques are needed to identify causal interdependencies between systems across different scales while overcoming various noise sources. Therefore, time-varying information flow among diverse physiological processes across scales must be identified. Robust optimisation algorithms must reconstruct or infer the structure and dynamics of complex interdependent networks.

With respect to network physiology, there is a serious lack of methodology. The inherent complexity of physiological systems and the problems that arise from network physiology are beyond the scope of the current state of the art in temporal networks. Current approaches to temporal networks do not account for the complex dynamics of network nodes (individual physiological systems) and for individual network links (coupling forms) which vary in time. The current formalism employed in temporal networks is not adequate for physiologic networks. There is no established analytic instrumentarium and theoretical framework suitable to probe networks comprising diverse systems with different output dynamics, operating on different time scales, and to quantify dynamic networks of organ interactions from continuous streams of noisy and transient signals.

From a medical point of view, there is currently no theory that connects the modelling of human physiological processes and the design and optimisation of healthcare systems. New machine learning and AI algorithms are requested to measure and mine the patient's physiological state based on available continuous sensing. Risk indices corresponding to the onset of abnormality must be identified to signal the need for critical medical intervention in real time by communicating patient's medical information via a network from an individual to the hospital. Such a network must include vital health signals (e.g. cardiac pacing, insulin level, blood pressure) within personalised homeostasis. Medical applications are heart dynamics and cancer patient's health [63].

Network physiology opens new perspectives investigating brain–brain network interactions across distinct brain rhythms and locations, characterising dynamical features of brain–organ communications as a new signature of neuroautonomic control, establishing basic principles underlying coordinated organ–organ communications, and constructing first dynamic maps of physiological systems and organ interactions across distinct physiologic states.

Network physiology shifts the focus from single organs to the network of physiologic interactions. Investigations in the field will explain how health emerges as a result of network interactions among systems. Coordinated interdisciplinary research efforts in the field will establish basic principles of organ integration essential to generate emergent behaviours at the organism level, and to facilitate responses and adaptation to internal and external perturbations, and thus, will redefine physiological states and functions in health and disease through unique network maps of physiologic interactions.

New mathematical and computational methods must be developed to address the complexity of physiological systems, to facilitate empirical findings of physiological interactions and to build the first theoretical framework for investigations of emerging global behaviours in networks of dynamical systems. This approach covers areas of applied mathematics, computer and data science, and network theory. It includes new techniques for physiological data analyses, and new network models of dynamical systems with time-dependent interactions to identify procedures of hierarchical integration, global network evolution across states and reorganisation between distinct network modules, motifs and communities of integrated physiological systems and subsystems.

Future developments in network physiology will generate new knowledge and understanding of complex systems concerning regulations and co-ordinations of organ-to-organ interactions, quantitative measures of the interactions between diverse organ systems and their collective network behaviour. Further on, physiologic complex systems are characterised by special relations between physiologic states and patterns of organ network interactions, hierarchical structure of physiological networks, procedures of network control and reorganisation with states, conditions and disease. Therefore, new areas of research are opened at the interface of computational and data science, applied mathematics and physics, AI and bioengineering, and physiology and medicine. They grow together in the new research field of network physiology, which integrates diverse scientific communities across a broad range of disciplines from applied mathematics, physics, data science and biomedical engineering to neuroscience, physiology and clinical medicine.

From complex dynamical systems to cellular neural networks

In general, we imagine a spatial system consisting of identical elements ('cells') that can interact with each other in different ways (e.g. physically, chemically or biologically). Such a system is called complex if it can generate non-homogeneous ('complex') patterns and structures from homogeneous initial conditions. This pattern and structure formation is triggered by local activity of its elements. This applies not only to, for example, stem cells during the growth of an embryo but also to transistors in electronic networks.

We call a transistor locally active when it can amplify a small signal input from the energy source of a battery to a larger signal output to generate non-homogeneous ('complex') voltage patterns in switching networks.

No radios, televisions or computers would be without the local activity of such units. Important researchers such as the Nobel Prize winners I. Prigogine (chemistry) and E. Schrödinger (physics) were still of the opinion that a non-linear system and an energy source are sufficient for structure and pattern formation. However, the example of transistors already shows that batteries and non-linear switching elements alone cannot generate complex patterns if the elements are not locally active in the sense of the described amplifier function.

The principle of local activity is of fundamental importance for the pattern formation of complex systems and has not yet been widely recognised. It can be defined mathematically in general, without having to rely on special examples from physics, chemistry, biology or technology. We refer to non-linear differential equations as known from reaction–diffusion processes (but by no means limited to liquid media as in chemical diffusion). We can clearly imagine a spatial lattice whose lattice points are occupied by cells that interact locally.

Every cell (e.g. protein in a cell, neuron in the brain, transistor in the computer) is mathematically a dynamic system with input and output. A cell state develops locally according to dynamic laws depending on the distribution of neighbouring cell states.

In summary, the dynamic laws are defined by the equations of the state of isolated cells and their coupling laws. In addition, initial and auxiliary conditions must be taken into account in the dynamics.

Definition of local activity

In general, a cell is called locally active if a small local input exists at a cellular equilibrium point that can be amplified with an external power source to a large output. The existence of an input that triggers local activity can be systematically tested mathematically by certain test criteria. A cell is called locally passive if there is no equilibrium point with local activity. What is fundamentally new about this approach is the proof that systems without locally active elements do not, in principle, have complex structures and are unable to create patterns.

Structure formation in nature and technology can be systematically classified by modelling application areas by reaction–diffusion equations according to the pattern described previously. For example, the corresponding differential equations for pattern formation in chemistry (e.g. pattern formation in homogeneous chemical media), in morphogenesis (e.g. pattern formation of mussel shells, fur and feathers in zoology), in brain research (circuit patterns in the brain) and in electronic network technology (e.g. circuit patterns in computers) can be investigated.

In statistical thermodynamics, the behaviour is determined by interaction of many elements (e.g. molecules) in a complex system. L. Boltzmann's 2nd law of thermodynamics only states that all structures, patterns and orders decay in an isolated system if one leaves them to oneself. Thus, all molecular arrangements dissolve in a gas, and heat is distributed uniformly and homogeneously in a closed space during dissipation. Organisms disintegrate and die if they are not in mass and energy exchange with their environment. But, how can order, structure and patterns be created?

> The principle of local activity explains how order and structure are created in an open system through dissipative interaction or mass and energy exchange with the system environment. It complements the 2nd law as the 3rd law of thermodynamics.

Structural formations correspond mathematically to non-homogeneous solutions of the considered differential equations, which depend on different control parameters (e.g. chemical concentrations, ATP energy in cells, neurochemical messengers of neurons). For the considered examples of differential equations, we could systematically define the parameter spaces whose points represent all possible control parameter values of the respective system. In these parameter spaces, the regions of local activity and local passivity can be precisely determined, which either enable structure formation or are mathematically 'dead'. In principle, computer simulations can be used to generate the possible structure and pattern formations for each point in the parameter space (Fig. 4). In this mathematical model framework, structure and pattern formation can be completely determined and predicted.

Local activity at the edge of chaos

A completely new application of local activity is the 'edge of chaos', where most complex structures arise. Originally stable ('dead') and isolated cells can be 'brought to life' by dissipative coupling and trigger pattern and structure formation. These cells can be imagined to 'rest' isolated at the edge of a stability zone until they become active through dissipative coupling.

One could imagine isolated chemical substances resting in the hostile dark deep sea at the edge of a hot volcanic vent. The dissipative interaction of the originally 'dead' elements leads to the formation of new forms of life. As chemical substances, however, they must carry within them the potential of local activity triggered by dissipative coupling.

This is unusual in that it seems to contradict the intuitive understanding of 'diffusion'. According to it, 'dissipation' means that, for example, a gas is distributed homogeneously in a space. However, not only unstable but also stable elements can trigger complex (inhomogeneous) structure and pattern formations by dissipative coupling. This can be proven exactly for non-linear reaction and diffusion equations. In the parameter spaces of these equations,

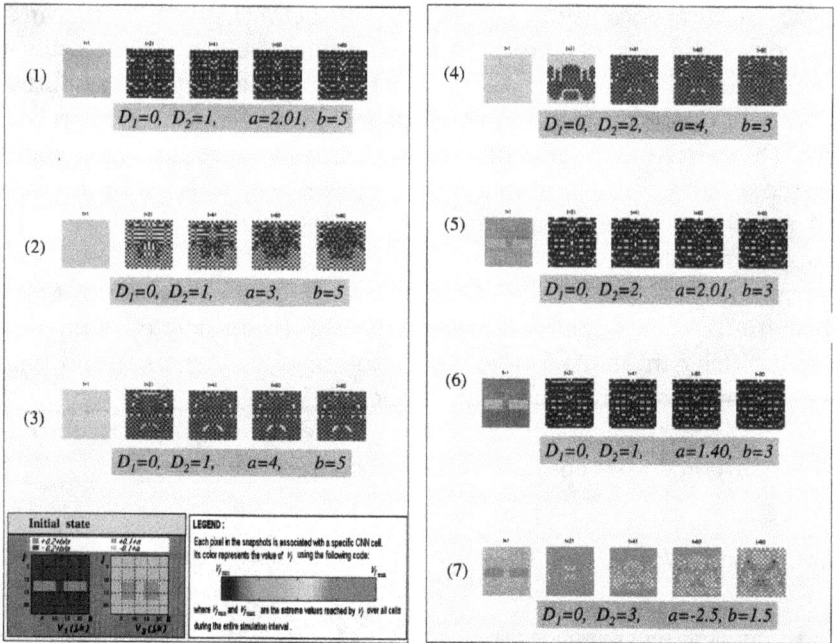

Fig. 4. Structure and pattern formation of a non-linear diffusion and reaction equation [64].

the 'edge of chaos' can be marked as part of the region of local activity.

Brains as complex dynamical systems

Even the human brain is an example of a complex dynamic system in which billions of neurons interact neurochemically. Complex switching patterns are created by multiple electrical impulses, which are associated with cognitive states such as thinking, feeling, perceiving or acting. The emergence of these mental states is again a typical example of self-organisation of a complex system: the single neuron is quasi 'stupid' and can neither think nor feel nor perceive. Only their collective interactions and interconnections under suitable conditions generate cognitive states.

The neurochemical dynamics between the neurons take place in the neuronal networks of brains. Chemical messengers cause neuronal state changes through direct and indirect transmission mechanisms of great plasticity. Different network states are stored in the synaptic connections of cellular switching patterns (cell assemblies). As is usual in a complex dynamic system, we also differentiate in the brain between the micro states of the elements (i.e. the digital states of 'firing' and 'non-firing' when a neuron is discharged and at rest) and the macro states of pattern formation (i.e. switching patterns of jointly activated neurons in a neural network).

Computer visualisations (e.g. PET (Positron Emission Tomography) images) show that different macroscopic circuit patterns are correlated with different mental and cognitive states such as perception, thinking, feeling and consciousness. In this sense, cognitive and mental states can be described as emergent properties of neural brain activity: individual neurons can neither see, feel, nor think, but brains connected to the sensors of the organism can.

Current computer simulations therefore observe pattern formation in the brain, which we attribute to non-linear system dynamics, the local activity of neurons and the action potentials they trigger. Their correlations with mental and cognitive states are revealed on the basis of psychological observations and measurements: whenever people see or speak this or that, pattern formation can be observed in the brain. In brain reading, individual patterns can now be determined to such an extent that the corresponding visual and auditory perceptions can be decoded from these circuit patterns using suitable algorithms. However, this technique is still in its infancy.

In a top-down strategy, neuropsychology and cognitive research are investigating mental and cognitive abilities such as perception, thinking, feeling and consciousness, and try to connect them with corresponding brain areas and their interconnection patterns. In a bottom-up strategy, neurochemistry and brain research investigate the molecular and cellular processes of brain dynamics and explain neuronal brain interconnection patterns, which in turn are correlated with mental and cognitive states [65].

Both methods suggest a comparison with the computer, in which, in a bottom-up strategy, the meanings of higher user languages of humans is derived from the 'machine language' of the bit states in, for example, transistors, while in a top-down strategy, conversely, the higher user languages are translated to the machine language via various intermediate stages (e.g. compiler and interpreter). However, while in computer science the individual technical and linguistic layers from the interconnection level via machine language, compiler, interpreter, etc. to the user level can be precisely identified, brain and cognitive research has so far only been a research programme.

In brain research, so far only the neurochemistry of neurons and synapses and the pattern formation of their circuits are well understood, i.e. the 'machine language' of the brain. The bridge (middleware) between cognition and 'machine language' has yet to be reconstructed. This will require many more detailed empirical studies. It is by no means already clear whether individual hierarchical levels can be precisely distinguished as in computer design. Apparently, the architecture of brain dynamics proves to be much more complex. In addition, the development of the brain was not based on a planned design but on a multitude of evolutionary algorithms that were developed more or less randomly under different conditions over millions of years and are connected to each other in a complex way.

Hodgkin–Huxley model of complex brain dynamics

In complexity research, the synaptic interaction of neurons in the brain can be described by coupled differential equations. The Hodgkin–Huxley equations are an example of non-linear reaction diffusion equations that can be used to model the transmission of nerve impulses. These equations have been struck down by the Nobel Prize winners in medicine A. L. Hodgkin and A. F. Huxley by empirical measurements and provide an empirically confirmed mathematical model of neuronal brain dynamics.

Axon

Neuron

(a)

dissipative (diffusion) coupling

Hodgkin Huxley cell

(b)

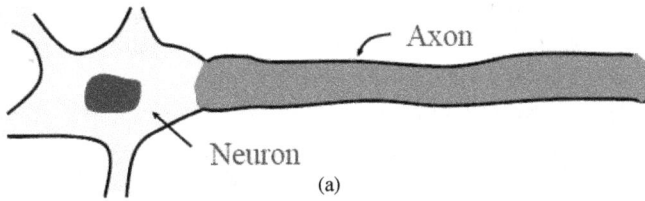

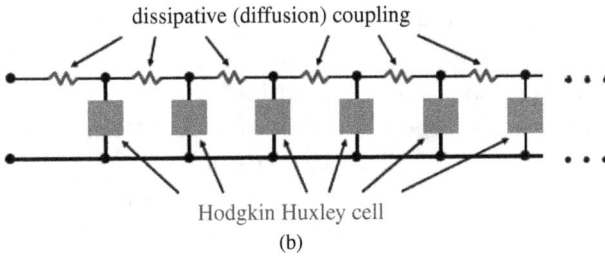

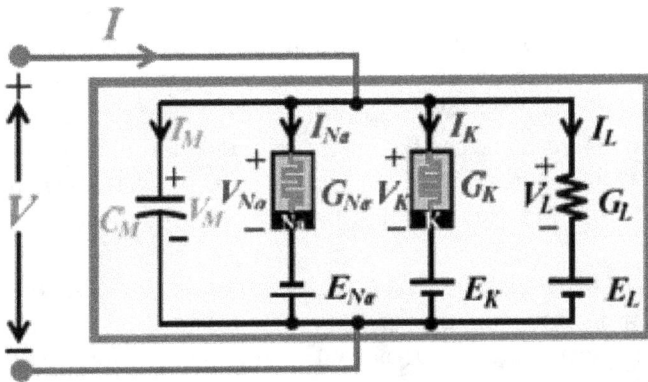

I external axon membrane current	E	membrane capacitor voltage
I_{Na} sodium ion current	E_{Na}	sodium ion battery voltage
I_K potassium ion current	E_K	potassium ion battery
I_L leakage current	E_L	leakage battery voltage

(c)

Fig. 5. Electrotechnical model of Hodgkin–Huxley equations.

In Fig. 5, the information channel (axon) of a neuron (a) is represented by a chain of identical Hodgkin–Huxley (HH) cells coupled by diffusion compounds (b). These couplings are technically represented by passive resistors. The HH cells correspond to an electrotechnical interconnection model (c): in a biological nerve cell, ionic currents of potassium and sodium alter the voltages on the cell membrane. In the electrotechnical model, sodium and potassium ion currents are triggered together with a current outflow by an external axon membrane current. The ion channels are technically realised by transistor-like amplifiers. They are connected to a sodium ion and potassium ion battery voltage, a membrane capacitor voltage and a voltage discharge. In this way, the input flows can be strengthened according to the principle of local activity in order to create a potential for action if a threshold value is exceeded ('fire'). These action potentials trigger chain reactions that lead to interconnection patterns of neurons.

As already explained, such differential equations can be used to precisely determine the corresponding parameter spaces of a dynamic system with locally active and locally passive regions. In the case of the Hodgkin–Huxley equations, we obtain the parameter space of the brain with the precisely measured regions of local activity and local passivity. Action potentials of neurons that trigger circuit patterns in the brain can only develop in the area of local activity. Computer simulations can be used to systematically investigate and predict these circuit patterns for the various parameter points.

Thus, the region at the 'edge of chaos' can be determined exactly. It is tiny and is less than 1 mV and 2 μA. This region is associated with large local activity and pattern formation, which can be visualised in the corresponding parameter spaces. Therefore, it is assumed that this is an 'island of creativity'.

For an electrotechnical realisation, however, the original equations of Hodgkin and Huxley proved to be faulty. The physicians Hodgkin and Huxley interpreted some switching elements in a way that led to electrotechnical anomalies. For example, they assumed a time-dependent conductivity (conductance) to explain the behaviour of the potassium and sodium ion channels. In fact, these temporal changes could only be calculated numerically from empirically derived equations. Theoretically, it was not possible to explicitly define corresponding time functions for time-varying switching elements.

The anomalies dissolve when the ion channels are explained by a new switching element that Leon Chua had already mathematically predicted in 1971 [66].

This refers to the memristor (from the English word 'memory' for memory and 'resistor' for resistor). With this switching element, the electrical resistance is not constant but depends on its past. The actual resistance of the memristor depends on how much charge has flowed in which direction. The resistance is maintained even without energy supply.

This realisation has enormous practical consequences but could also be a breakthrough for neuromorphic computers oriented to the human brain. First, we explain the concept of a memristor.

In practice, computers equipped with memristors would be ready for operation immediately after switching on without booting. A memristor retains its memory content if it is read with alternating current. A computer could therefore be switched on and off like a light switch without information being lost.

Classical digital computer versus neuromorphic computer

The computer architecture of a classic digital computer is still attributed to the Hungarian-American mathematician John von Neumann, which was used in one of the first universally programmable digital American computer (e.g. ENIAC 1945). Digital information processing requires electrons to be moved between a working memory, computing and control units. Although this is a clear order on which programming languages can be logically based, from today's point of view it generates a serious and environmentally damaging amount of energy.

Information is coded as bit sequences of 0 and 1 and is constantly changed in calculation steps. All computing steps must also be processed one after the other (sequentially). The decisive factor is that bits have to be moved back and forth between the separate memory, computing and control units for each computing step. With ever smaller chip structures and growing amounts of information (big data), this creates a bottleneck in information processing, which is also known as the 'von Neumann bottleneck' with regard to computer architecture. This term goes back to the developer of the programming language 'Fortran', John W. Backus (1977).

However, it is not only the increasing computing time that the 'von Neumann bottleneck' generates through the sequential digital processing of large amounts of data. The movement of electrons also requires energy and generates heat. This waste heat increases with ever smaller chip structures. If neural networks with hundreds and thousands of neural layers and gigantically increasing numbers of parameters are now to be simulated on von Neumann computers in modern deep learning, then you can imagine how the von Neumann bottleneck will become an 'energy guzzler' whose CO_2 emissions will exacerbate the climate problem.

A paradigm shift in computer structures is unavoidable if the growing use of AI, applications in mobile phones, autonomous vehicles, medical technology and sensor networks is to be realised in a resource-efficient manner. This is the only way to achieve sustainable computer technology that fulfils the 'Green Deal' and is also economically effective, i.e. lower costs with a simultaneous increase in the performance of AI applications. This is where neuromorphic computer structures modelled on natural brains come in.

*Foundations of memristive systems**

Traditionally, in electrical engineering, only resistance, capacitor and coil were distinguished as switching elements. They combine the four switching variables, namely, charge, current, voltage and magnetic flux: resistors connect charge and current, coils connect magnetic flux and current, capacitors connect voltage and charge. But what connects charge and magnetic flux? L. Chua postulated the memristor in 1971. Mathematically, this is done with a function $R(q)$ (memristance function) in which the change of the magnetic flux Φ with charge q is being held, i.e.

$$R(q) = \frac{d\Phi(q)}{dq}.$$

The temporal change of the charge q defines the current $i(t)$, i.e.

$$i(t) = \frac{dq}{dt}.$$

The temporal change of the magnetic flux Φ defines the voltage $v(t)$, i.e.

$$v(t) = \frac{d\Phi}{dt}.$$

The result is that the voltage v at a memristor about the current i depends directly on the memristance:

$$v = R(q)\,i.$$

This is reminiscent of Ohm's law $v = Ri$, which defines that the voltage v is proportional to current i with the resistance R as a proportionality constant. However, the memristance is not constant, but depends on the state of the charge q. Conversely, for electricity, it holds

$$i = G(q)v$$

where the function $G(q) = R(q)^{-1}$ is called 'memductance' (composed of the English word 'memory' for memory and 'conductance' for conductivity).

A memristor can be generalised as a memristive system. A memristive system is no longer reduced to a single state variable and a linear charge- or flow-driven equation.

A memristive system is an arbitrary physical system, which is defined by a set of internal state variables $\vec{s}$ (as a vector). It follows a general input–output equation

$$\vec{y}(t) = g(\vec{s}, \vec{u}, t)\,\vec{u}(t)$$

with the input $\vec{u}(t)$ (e.g. voltage) and the output $\vec{y}(t)$ (e.g. electricity). The development of a state is generally determined by a differential equation:

$$\frac{d\vec{s}}{dt} = f(\vec{s}, \vec{u}, t).$$

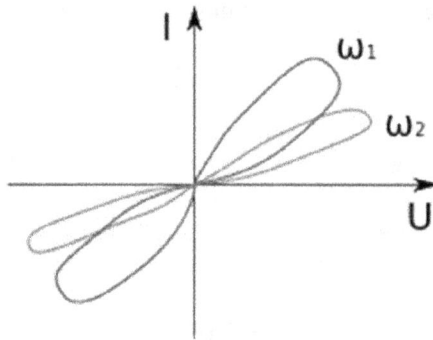

Fig. 6. Hysteresis curve of a memristor (depending on the angular frequency) ω with $\omega_1 < \omega_2$).

Memristive systems exhibit exceptionally complex and non-linear behaviour. Typical is the hysteresis curve in the v/i — diagram in Fig. 6. It runs in closed loops through the pinched hysteresis loop [67].

In general, hysteresis refers to the behaviour of the output variable of a system that reacts to an input variable with a delayed (Greek hysteros) signal and varies. The behaviour does not only depend directly on the input variable but also on the previous state of the output variable. If the input variable is the same, the system can adopt one of several possible states.

As neuristors, memristive systems simulate the behaviour of synapses and therefore become interesting for neuromorphic computers. For this purpose, the ion channels in the circuit model in Fig. 5 are regarded as memristive systems. Hodgkin–Huxley's time-dependent conductivity G_k of the potassium ion channel is replaced by a charge-controlled memristor dependent on a state variable. Further, Hodgkin–Huxley's time-dependent conductivity G_{Na} of the sodium ion channel is replaced by a charge-controlled memristor which depends on two state variables. These well-defined quantities explain precisely the empirical measurement and observation data of synapses and neurons [68].

But how can such neuristors be technically realised? R. Stanley Williams of the company Hewlett Packard (Silicon Valley) constructed a version for the first time in 2007, which has meanwhile been constantly simplified and improved [69]. Imagine a crossbar

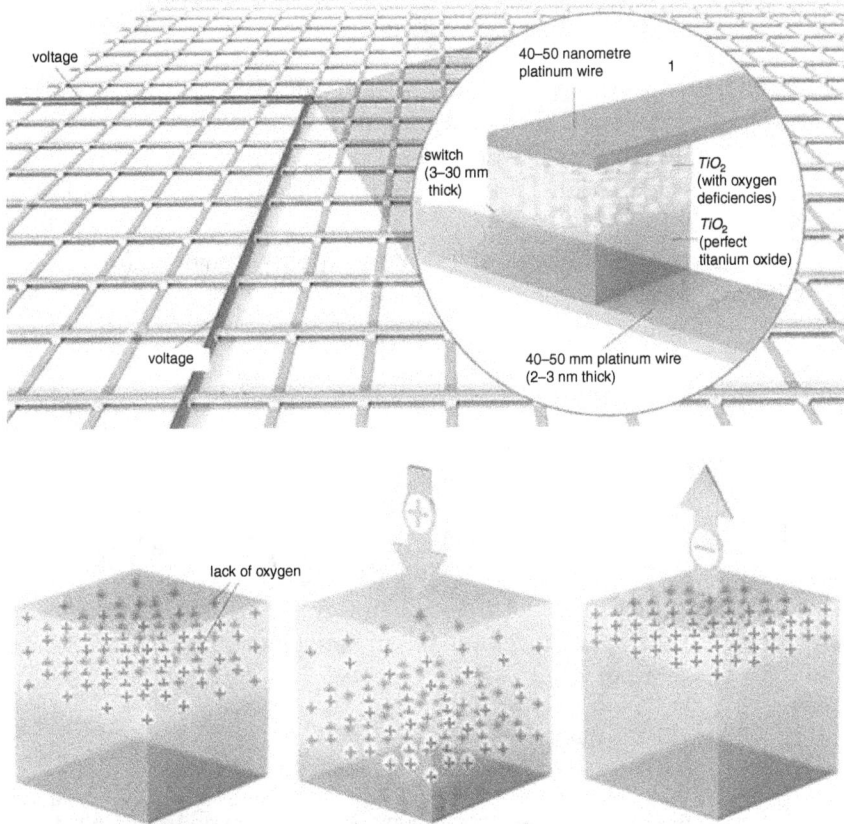

Fig. 7. Memristive system with switches made of titanium dioxide.

network of intersecting vertical and horizontal wires, reminiscent of a wire mesh (Fig. 7) [70]. The intersections of a vertical and horizontal wire are connected by a switch. To close the switch, a positive voltage is applied to both wires. To open it, the charge is reversed.

In order to achieve memristive behaviour, the switches are constructed according to a specific architecture. It is reminiscent of a sandwich in which a titanium dioxide layer a few nanometres thick lies between two platinum electrodes (as 'slices of bread'). In Fig. 7, the lower titanium dioxide layer serves as an insulator. The upper titanium dioxide layer has oxygen deficiencies. You can imagine them as small bubbles in a beer — with the difference that they cannot escape. This titanium oxide layer has a high conductivity. If a positive voltage is applied, the oxygen deficiencies shift. This reduces the thickness of the lower insulation layer and increases the overall conductivity of the switch. A negative charge, on the other hand, attracts the positively charged oxygen deficiencies. This increases the insulation layer and reduces the overall conductivity of the switch.

The memristive behaviour becomes apparent when the voltage is switched positively or negatively: then the small bubbles of the oxygen deficiencies do not change but remain where they are. The border between the two titanium dioxide layers is 'frozen', so the switch can 'remember' how much voltage was last applied. It works like a memristor.

Other memristors use silicon dioxide layers a few nanometres in size, which require only low costs. The crossbar memories manufactured by Hewlett Packard already have an enormous packing density of approx. 100 Gibit/cm^2. They could also be combined with other semiconductor structures. Therefore, it cannot be excluded that they initiate the development of neuromorphic structures to simulate the human brain.

Cognition through memristive networks

The outcome of this research programme was the mathematical Hodgkin–Huxley model of the brain. In the Human Brain Project of the European Union, an exact empirical modelling of the human brain with all neurological details is aimed at. With the technical development of neuromorphic networks, an empirical test bed would be available for this mathematical model in which predictions about pattern formation in the brain and their cognitive meanings can be verified.

From psychology, we know that mental and cognitive states interact in an extremely complex way. Perceptions can thus trigger thoughts and ideas that lead to actions and movements. However, a perception is usually also connected with a self-perception: it's me who perceives. Self-awareness, combined with the storage of one's own biography in memory, leads to ego-consciousness. If all these different mental states are associated with circuit patterns in the brain, then not only the interactions of individual neurons must be recorded but also those of cell assemblies with cell assemblies of cell assemblies and so on.

In principle, differential equations can also be introduced that do not depend on the local activities of individual neurons but on whole cell assemblies, which in turn can depend on cell assemblies of cell assemblies, and so on. This results in a system of non-linear differential equations which are interconnected on different levels and thus model an extremely complex dynamic. Connected to the sensors and actuators of our organism, they record the processes that create our complex motor, cognitive and mental states. As already stressed, we do not yet know all these processes in detail. But it is clear how, in principle, they can be mathematically modelled and empirically tested in neuromorphic computers.

Memristor crossbar architectures

Electronic components whose resistance changes as a result of electrical stimuli such as current and voltage are called memristive switching elements or memristors [71]. The variants of memristive switching elements are based on different physical principles such as defect-based resistive memories (ReRAM), phase change (PCM) and magnetic as well as ferroelectric tunnel junctions. These devices were initially applied as non-volatile memories (NVM) due to their scalability, silicon compatibility and performance advantages compared to standard NVM devices.

In artificial neural networks, the synaptic weights between neurons can be simulated by memristors. One speaks of memristor crossbar arrays when the complete connection (connectivity) between two neuron levels is mapped into a two-dimensional array with memristors at the connection points [72]. The arrays enable highly energy-efficient, fully parallel, analogue in-memory computation of vector-matrix products. They avoid the computational bottleneck ('von Neumann bottleneck') during artificial neural network training using standard hardware with central processing units (CPU) or graphics processing units (GPU).

Therefore, memristive crossbar arrays are a future-oriented hardware technology for deep learning AI neural networks. Memristive devices can replicate brain-like synaptic processes. This includes processes such as Spike-Timing-Dependent Plasticity (STDP) as an important form of local learning rules to enable self-learning of brain-like neuromorphic systems.

Photonic neuromorphic circuits

An alternative to electronic data processing is light [73]. Photons as light particles are 1000 times faster compared to electrons in circuits. In addition, light waves do not influence each other and have a low energy requirement. The photonics based on this developed on the basis of light emitters with semiconductors and optical fibres. Integrated photonic components therefore open up the possibility of combining high speeds, parallel data streams and low consumption.

Optical transmission systems enable high data rates and a reduction in latency during signal transmission. Complex systems require sufficient scalability of the number of neurons and synapses. Here, a hybrid solution is offered that implements signal processing optically and electronically.

The non-linear transfer functions of the neurons and the signal regeneration are carried out electrically, while the signal transmission between the neurons and the linear signal processing during the weighting and summation of the signals are realised optically. In this way, the delay times can be decisively reduced. A proof-of-concept demonstrator of a corresponding neural network is a decisive innovation step for long-term goals of serial production.

Conventional silicon technology, which can draw on long experience of digitisation in fabrication, is also used. Neuromorphic principles are to be taken into account in the development of novel algorithms and component properties. To this end, insights from neuroscience, automated system design and hardware-related circuit development are to be brought together. The innovation aims at circuit architectures that integrate electrical processing and photonics.

Energy consumption of brains compared with memristive and photonic systems

To measure the energy consumption per computing unit, a multiply–accumulate (MAC) operation is used [74]. This involves multiplying two factors and adding the product to a continuous summand. MAC operations are also used to represent the course of synaptic functions. The computational efficiency of the human brain is 20 W for 10^{18} MAC per second. This assumes 10^{11} neurons with about 10,000 connections of a neuron to other neurons. This gives a total of 10^{15} synaptic connections, all transmitting signals with a bandwidth of up to 1 kHz. The computational efficiency of the brain is thus less than one attojoule (= 10^{-18} joules) per MAC in contrast to about 1 picojoule (= 10^{-12} joules) per MAC in a powerful computer.

Neuromorphic systems based on analogue electronics with memristors or photonic chips open up significant improvements in computing power. Although the amplitude of the spikes is analogue in time, it is represented digitally. An example of higher performance in neuromorphic processing is high-bandwidth applications in the GHz range. These include scanning and manipulating the radio spectrum or controlling fast aircraft based on photonic chips.

Electronic circuit systems require billions of switches that change between an on and off state. This process leads to overall time delays (latency). In contrast, photonics relies on wave propagations with interference patterns to determine the result. This allows direct calculations without delays due to switch latency. Photonics achieves high switching speeds and communication bandwidths with little interference from neighbouring communication channels. This allows very fast information processing based on spikes with high link density. It is anticipated that photonics-based neuromorphic systems could work up to 100 million times faster than neuromorphic electronics. This will definitely achieve speeds like biological brains and even faster.

Energy consumption with photonic systems also opens up remarkable perspectives. Theoretically, optical neural networks could avoid a physical limit of traditional digital chips. Thermodynamically, these chips work irreversibly. This limit became known as the Landauer limit, according to which the erasure of even a small amount of information releases a minimal amount of energy as heat to the environment. However, there is a lower limit to this amount. For photonic circuits, the limit depends on the network and the problem being worked on. In tested examples, it is between 50 zJ to 5 aJ/MAC. The thermodynamic Landauer limit for a digital processor is 3 aJ/MAC for 1000-bit operations per MAC.

Hybrid AI systems as sustainable innovation

Each technology has advantages and disadvantages that need to be weighed up against each other. In addition, technologies prove to be bridges to new and more efficient solutions. They are therefore always also 'bridging technologies'. In such a situation, it would be extremely unwise to focus solely on an existing technology or to hope for the future. As with shares, it is important to network a portfolio of different technologies in order to make a good cut in the event of failures and misjudgements of individual parts. Such an innovation portfolio is also subject to dynamic change and must be maintained and cultivated.

Hybrid AI = symbolic AI + subsymbolic AI

As shown in the previous chapter, a dynamic innovation portfolio is necessary for energy technologies and also for computer technologies and the information and communication systems with artificial intelligence that build on them. With regard to artificial intelligence, it has already been emphasised that modern machine and deep learning will by no means 'disruptively' replace traditional AI. Rather, it is important to combine symbolic AI as a control and verification process with the statistical methods of subsymbolic AI. Subsymbolic AI is primarily aimed at pattern recognition, as occurs in organisms in perceptual processes, while symbolic AI maps the logical thinking and reasoning of the human mind (e.g. in expert systems). The connection of both concepts is called hybrid AI. It approximates human intelligence more than the reduction to only one of them [75].

Hybrid computing = classical computing + quantum computing

Traditional computing, from smartphones and PCs to supercomputers, also has a hybrid relationship with quantum computing. When it is fully developed, quantum computing will not 'disruptively' replace conventional computing. Rather, the functions of quantum computing will be embedded in classic mainframe computer systems in order to solve typical tasks in combination with classic procedures. This also applies to the existing D-Wave computers, which are not based on circuits with quantum logic gates like quantum computers but solve optimisation tasks using adiabatic computing as in thermodynamics. The difference to classical adiabatic computing is that thousands of quantum bits are already used, albeit simulated on conventional mainframe computers. It is therefore again a question of solving special tasks ('optimisation tasks') using special methods ('adiabatic computing with quantum bits') that are embedded in conventional computer systems.

Hybrid computer = analogue computer + digital computer

Before universal programmable digital computers laid the foundation for modern digitisation in the 1940s, analogue computers were used, which were also designed to solve special tasks. The idea of analogue computers is an ancient one [76]: for example, one builds a technical–mechanical model ('analogue') of the physical planetary system in order to use it to determine constellations of the planets. As early as the 1st century BC, Greek mathematicians presented such an analogue computer with the Antikythera mechanism, which was able to perform such calculations by adjusting a cogwheel mechanism for assumed celestial spheres. With the advent of electrical engineering, tubes, transistors, etc. could be used to realise computational tasks. In a differential or integral equation, the calculation operators, for example, multiplication, addition, integration, differentiation, were represented by corresponding ('analogue') electrotechnical units and technically connected with each other to solve the task. The mathematical equation can be understood as a circuit diagram for the analogue technical–physical model with which the solutions are calculated.

An analogue computer therefore requires a large number of arithmetic elements to represent all the different operations, variables and links in tasks in 'analogue' form. In contrast, a digital computer requires only a few arithmetic units, which have to be processed extremely quickly with very simple instructions one after the other. These are the bits 0 and 1 with a few operations of adding, subtracting and multiplying bits, which can be linked in a few logical gates to form highly complex circuits. For an addition task, a digital computer must first read the bit sequence for the 'Add' command from a memory and execute it. The summands must also be read from a memory as bit sequences before the operation is carried out on the summands. All this takes time and energy, which adds up to millions and millions of calculation steps for individual bits. An analogue computer does not require a von Neumann architecture with separate memory, control and computing units and sequential execution of instructions in algorithms and programs but solves the tasks 'directly' in an electronic model.

From today's perspective, analogue computers are therefore energy-efficient and time-saving, but are limited to special applications that can be represented in differential and integral equations, for example. These are primarily special tasks from engineering, natural and physical sciences for which an analogue computer model must be created. Mathematically, such equations use real numbers such as decimal fractions (e.g. $\pi = 3.1415 \ldots$ with arbitrary ('infinitely small') quantities and continuous processes in contrast to digital numbers such as the two bits. Analogue computing therefore also refers to computing with real numbers [77].

In the 1970s, it initially seemed that digital computers would replace analogue computers due to their universal applicability in all possible areas. Due to the tremendous speed of digital computers, from smartphones to supercomputers, it seemed that the typical tasks of analogue computers could also be simulated. From the point of view of energy efficiency and environmental protection, analogue computers are now once again becoming the focus of interest.

Hybrid computers are now also understood to mean the coupling of digital and analogue computers [78]. The analogue computer is used as a powerful co-processor for the digital computer.

However, this also requires software to be able to program the analogue computer in the computer environment of the digital computer. Similar to the functions of a quantum computer, analogue computers are embedded in a digital environment for solving special tasks. These are also referred to as 'ecological' computer systems and refer to the diverse networking of computer types.

Hybrid robotics = embodied robotics ('embodied mind') = analogue robotics + digital robotics

In robotics, too, digital and analogue functions are combined as in an organism. Cognitive and intellectual abilities cannot be modelled in software that is separate from the body. In order to recognise connections and structures, experience is required through perception processes with sensory organs and the experience of movement sequences with organic motor skills such as hands and fingers.

In cognitive psychology and cognitive philosophy, the term 'embodied mind' is used: the human 'mind' is not isolated but 'embodied' in the organism. Many physical processes take place in an analogue manner via sensors, while control and steering functions tend to be digital. Humanoid robots are increasingly being developed along the lines of the 'embodied mind' in a hybrid coupling of analogue and digital functions [79].

Analogue functions modelled on the brain also come into play in neuromorphic computing. Biological brains are by no means digital switch boxes in which neurons 'fire' and 'do not fire', i.e. digitally switch back and forth between two states as bits 0 and 1. Rather, synaptic connections have analogue components, as they take continuous and gradual weightings into account. As explained in section 'Quantum AI systems', analogue and digital components are connected in memristive and photonic neural systems. However, the hardware of neuromorphic systems is realised with silicon and nanotechnological materials and not with living tissue ('wetware') as in biological brains. On the one hand, this hardware is certainly more robust and cannot be affected by diseases like biological tissue. It is also possible to increase performance and withstand continuous stress, which living tissue cannot withstand, including the fatigue of natural brains.

Neurobiological hybrid brains

However, a decisive disadvantage of the hardware of neuromorphic systems to date is the fact that they cannot change on their own and develop like biological brains. The biological brain of a child grows, but the hardware of a neuromorphic system does not. To this end, genes encode the growth process that leads to the networking of living neuronal cells. This growth process depends on numerous interactions and is subject to the non-linear dynamics of a complex system [80]. As a result, cognitive functions and intelligence, emotions and states of consciousness also emerge and develop and are not predetermined. Developmental psychology investigates the gene-controlled phase transitions of these growth processes on which the development of cognitive abilities such as learning depend.

Artificial neural networks, on the other hand, are technically pre-constructed. They either simulate cognitive abilities on digital computers, as in machine learning, or realise them with the hardware of neuromorphic systems. The 'wiring' of the artificial brains has thus far been predetermined in software or hardware. However, as these artificial networks are highly complex systems, an enormous amount of information is required to describe them in detail. This is also referred to as a parameter explosion in the deep learning of artificial neural networks.

In order to record the growth of biological brains, a much smaller amount of information is sufficient to record the genomes and genes that control the growth process [81]. Growth is time- and energy-dependent. The more time and energy is spent on growth, the more information can already be stored in a network before it begins to learn. This is why the brains of baby animals contain information that does not have to be learnt first. Like young foals, for example, they can stand immediately after birth and find their mothers' milk teats. There is no time for time-consuming training of neural networks through many examples with large amounts of data in the struggle for evolutionary survival.

Until now, the design of neural networks has been limited to simplifying abstractions that disregard biological details. Neuromorphic systems are already increasingly taking into account neurobiological details of synapses and neurones. In order to achieve the energy efficiency and performance of living brains, the results of developmental biology and developmental psychology will increasingly be taken into account. Then we will probably also discover the astonishing ability of living brains, which by no means require 'big data' to develop creative innovations. In the history of innovation, these brains have often made do with little information in order to develop innovations with previously misunderstood creative abilities.

Hybrid AI — Systems coupling humans and computers

People and their creative abilities must not be forgotten in the dynamic innovation portfolio. Hybrid IT systems therefore also refer to the coupling of algorithms and humans. While algorithms can process huge amounts of information in a flash with extreme precision, humans are generally more creative and are characterised by both intuition and empathy in dealing with their environment. Ultimately, decisions should still be made by humans. However, if an experienced pathologist can draw on thousands of examples when assessing tumours in tissue sections while deep learning algorithms can access millions of examples at lightning speed, abilities of human experts are clearly overcome by machines.

A human expert should not only be able to give a diagnosis but also to justify explanations and considerations. This explanatory component must also be demanded of algorithms (explainable AI). In the example of medicine, such explanations must ultimately stand up in court. This is where the limits of an algorithm become clear if there is no legal background knowledge. For this reason, the final decision and the associated responsibility for people in high-risk and explosive situations such as in medicine are mandatory.

Information exchange between humans and machines (algorithms) is required in hybrid systems. People learn from the machine and vice versa. This creates feedback loops of learning between two sides that complement each other. The example of medicine makes it clear that the final decision, e.g. on the application of a therapy for cancer, must not only take into account questions of precise tumour determination and possible feedback with other diseases of the patient but also psychological, social, legal and economic factors of quality of life, which are not (yet?) feasible for an algorithm. Even if it were possible to record all these factors and include them in an overall assessment according to an algorithm, this algorithm itself would have to be scrutinised. Therefore, the ultimate responsibility in this case would lie with the doctor.

Goal: Hybrid AI systems as a dynamic research portfolio

The goal is therefore a dynamic innovation portfolio of current IT systems that are hybridised with each other. The various bridging technologies can complement, reinforce or replace each other. Ultimately, however, they should be a sustainable service for us humans and this planet, also in terms of energy consumption and environmental impact.

It would be advisable if there could be a sustainability label for IT systems, for example, in the form of standards, as they are known from technology. An AI standardisation roadmap has already been proposed for Germany, in which all technical, economic, ecological, social, legal and ethical sustainability factors are included [82]. It must be made clear that this type of sustainability requirement does not act as a brake on innovation but rather promotes innovation as it provides legal certainty and guidance.

Goal: Hybrid AI as a hardware-independent service system for everyone

The current hardware development of neuromorphic computing can be compared with the period of digital computing since Turing's theoretical concept of a universal Turing machine from 1936 up to the first technical constructions with von Neumann architecture in the 1940s. What was missed since the end of the 1940s was the development of software languages to make computing machines applicable for common users beyond highly specialised knowledge of mathematicians and engineers. Concerning neuromorphic computing, the lack of programming models for already existing neuromorphic hardware has been obvious since many years.

With the Loihi 2 architecture of Intel (named after a volcano in the south of Hawaii), a new open-source software has come up, called Lava, for developing neuro-inspired applications [83]. Lava is modular and composable. Such as in digital computing, levels of abstraction from the hardware chips can be built up to make neuromorphic computing accessible to different applications. A low-level interface for executing neural network models is called Magma. It supports cross-platform execution, before being applied to Loihi 2 neuromorphic platforms. Lava specifies, compiles and executes processes which can be mapped to different platforms with conventional as well as neuromorphic components. It includes offline training and interfaces with applications to heterogeneous systems and real-world applications. All libraries, tools and features are exposed through the programming language Python with accelerated performance using underlying C/C++/CUDA code.

The fundamental building blocks in the Lava architecture are processes from which all algorithms and applications are constructed. Processes are objects with internal state variables, input and output ports for messages which are communicated via channels and behavioural models. Lava processes can be compiled and executed via a cross-platform, compiler and runtime which can be executed on neuromorphic as well as conventional von Neumann hardware. These components determine the low-level layer of Lava that is called Magma. For all higher levels the process library contains a growing set of new processes for neuron models, neural network topologies, Input–Output (IO) processes and all kinds of neuromorphic components.

As in digital computing, an operating system is needed which connects neuromorphic hardware with a software stack up to high-level languages of users without understanding the hardware machinery. An expert should, of course, be able to analyse the layers of software at any depth. Such as in digital computing, the software stack consists of several layers of abstraction. The basic layer is the neuromorphic hardware which can be realised by different materials. In any case, the materials should have memristive or photonic properties. Memristive materials such as, for example, titanium dioxide show a 'memorising' ability illustrated by a non-linear hysteresis curve in a voltage (v)/current (i)-diagram. The non-linear hysteresis curve is a signature for materials which are possible candidates to realise the behaviour of a synapse in a neuromorphic chip.

The neuromorphic hardware is followed by physical and software communication layers. In a next step, hardware abstraction and functionalities like mapping and routing must be realised. Next comes expert-level software, which is used in the last layer of user-level applications [84]. So, at the end, the user is able to describe and execute a neuromorphic experiment without knowledge of the underlying parts. This is useful on platforms of neuroscientists or physicians who do experiments in brain research or diagnosis and surgery in neurology and medicine.

Also in quantum computing, hardware-independent applications are the goal of future Internet applications. Up to now, only quantum network applications and functionalities on quantum processors have been realised in specialised software to restricted experimental constraints in a single task. Such as in classical digital and neuromorphic computing, we aim at a user platform with independent high-level software which needs no knowledge of the underlying technology, in this case quantum mechanics and quantum technology. Nowadays, quantum hardware is realised in different versions with different advantages and disadvantages. So far, here is now industrial standardisation such as in the early days of classical digital computing.

For example, a quantum bit can be realised by a photon with linear (vertical and horizontal) polarisation for the alternative states 0 and 1 and probabilistic scaling of intermediate states between 0 and 1 for more or less being 0 or 1. But also circular polarisation is used for photons, spins of electrons, neutrons or nuclei, internal states of atoms or ions, and energy levels of quantum dots. Therefore, a quantum network operating system is necessary with interfaces to different quantum hardware prototypes.

A quantum network application consists of several programs [85]. Each program runs on an end node which is a quantum device executing user applications. A network stack enables entanglement generation between nodes over a quantum network. A first quantum network operating system QNodeOS was introduced with nodes realised by a trapped-ion based on a single $^{40}Ca^+$ atom and nitrogen-vacancy (NV) centres in diamond (Fig. 8). This architecture is not only a first initiative in quantum network programming,

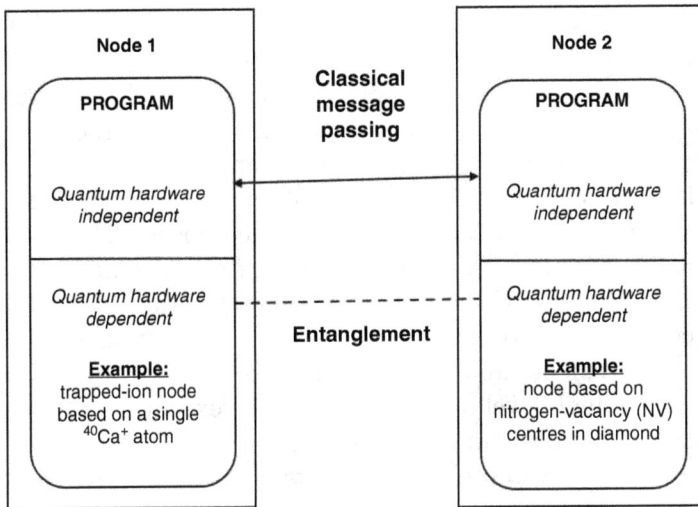

Fig. 8. Quantum operating system at the interface of quantum hardware-independent user languages and different quantum-hardware technologies.

the development of hardware-independent software opens avenues of quantum computing for scientific and societal applications. The operating system QNodeOS allows programs written in high-level quantum hardware-independent software and executed on a quantum hardware-independent system that controls different hardware-dependent quantum systems. Examples are NV centres with a diamond chip or a trapped-ion quantum node.

The programs are divided into classical and quantum blocks of instructions by a programmer or compiler. Classical blocks include local classical operations for classical processors. Quantum blocks include local quantum operations such as quantum gates, measurements and entanglement executed on quantum hardware. For the application of the operating system to the different quantum hardware, a hardware abstraction layer is necessary that is capable of interfacing with different quantum network processors.

The different parts of the quantum operating system operate at different time scales. For nodes at a distance of hundreds of kilometres, the network operations take milleseconds (ms). Quantum operations on processing nodes take microseconds (μs). The low level of neighbouring nodes to generate entanglement requires nanoseconds (ns).

The quantum operating system consists of a classical network processing unit (CNPU) and a quantum network processing unit (QNPU). The QNPU controls a quantum hardware device (QDevise). The CNPU is the logical element which starts the execution of programs and classical code blocks. The QNPU is the logical unit governing the execution of the quantum blocks. The CNPU and the QNPU determine the operating system QNodeOS, which controls the QDevice with quantum operations of quantum gates, measurements and entanglement on the quantum hardware. In this way, high-level programming and execution of quantum network applications can be connected.

The goal is a comprehensive software stack, in order to make quantum computing accessible to a wide range of users from different fields. The software track should be applicable to different hardware platforms and support all kinds of quantum algorithms. Even users unfamiliar with quantum physics should be able to use quantum computing resources. Quantum software should be integrated with existing software stacks for classical high-performance computing (HPC).

In the end, neuromorphic and quantum computing should be integrated in common platforms of users who are not familiar with neuroscience or quantum physics such as in the case of users of classical digital computing who have no knowledge of classical electrotechnology. In a supercomputing centre of the future, the user will find an integrated platform for all these different hardware applications. They are useful for different fields with different requests. For realisation of this integrating task, compilers and operating systems are necessary to translate neuromorphic and quantum algorithms to real neuromorphic and quantum systems in hardware. This kind of technological innovation needs interdisciplinary research integrating computer science, neuroscience, quantum physics, engineering and material science. But the user of the future need not have any idea of these complex systems, such as an animal or human who simply enjoys the abilities of his, her or its organism.

References

[1] Mainzer, K. (2019), *Artificial Intelligence. When Do Machines Take Over?* 2nd edn., Springer: Berlin, p. 3.

[2] Puppe, F. (1988), *Introduction to Expert Systems*, Springer: Berlin.

[3] Mainzer, K. (1990), Knowledge-based systems. Remarks on the philosophy of technology and artificial intelligence, *Journal for General Philosophy of Science*, 21, pp. 47–74.

[4] Vapnik, V. N. (1998), *Statistical Learning Theory*, Wiley: New York.

[5] Hornik, K., Stinchcombe, M., and White, H. (1989), Multilayer feedforward networks are universal approximators neural networks, *Neural Networks*, 2, pp. 359–366.

[6] Mainzer, K. (2024), Verifikation und Sicherheit für Neuronale Netze und Machine Learning, in Mainzer, K. (Ed.) *Philosophisches Handbuch der Künstlichen Intelligenz*, Springer: Berlin.

[7] Biere, A., Heule, M., van Maaren, H., and Walsh, T. (Eds.) (2009), *Handbook of Satisfiability*, IOS Press: Amsterdam.

[8] Pulina, L. and Taccella, A. (2012), Challenging SMT solvers to verify neural networks, *AI Communications*, 25(2), pp. 117–135.

[9] Leofante, F., Narodytska, N., Pulina, L., and Tacchella, A. (2018), Automated verification of neural networks: Advance, challenges and perspectives, *arXiv*: 1805.009938v1 [cs.AI].

[10] Ehlers, R. (2017), Formal verification of piece-wise linear feed-forward neural networks, *arXiv*: 1705.01320v3 [cs.LO].

[11] Mainzer, K. (2007), *Thinking in Complexity. The Computational Dynamics of Matter, Mind, and Mankind,* 5th edn., Springer: Berlin.

[12] Roberts, D. A., Yaida, S., and Hanin, B. (2022), *The Principles of Deep Learning Theory. An Effective Theory Approach to Understanding Neural Networks*, Cambridge University Press: Cambridge, p. 5.

[13] Roberts, D. A., Yaida, S., and Hanin, B. (2022), *The Principles of Deep Learning Theory. An Effective Theory Approach to Understanding Neural Networks*, Cambridge University Press: Cambridge, p. 9 ff, 391–399.

[14] Pearl, J. (2009), *Causality: Models, Reasoning, and Inference*, Cambridge, Mass, MIT Press.

[15] Hume, D. (1993), *Eine Untersuchung über den menschlichen Verstand*, 12th edn., in Übersetzt von R. Richter (Ed.) J. Kulenkampff, Hamburg, p. 95.

[16] Kant, I. (1900ff), *Werke. Ed. vols. 1–22 Preussische Akademie der Wissenschaften*. Bd. 23 Deutsche Akademie der Wissenschaften zu Berlin (ab Bd. 24 Akademie der Wissenschaften zu Göttingen), Berlin AA III, S. 93–KrV B 106.

[17] Knight, W. (2017), The dark secret at the heart of AI, *MIT Technology Review*, April 11, S. 1–22.

[18] Mainzer, K. (2020), Statistisches und kausales Lernen im Machine Learning, in Mainzer, K. (Ed.) *Philosophisches Handbuch der künstlichen Intelligenz*, Springer: Berlin.

[19] Mainzer, K. (2020), *Leben als Maschine: Wie entschlüsseln wir den Corona-Code? Von der Systembiologie und Bioinformatik zur Robotik und Künstlichen Intelligenz*, 2nd edn., Brill Mentis: Paderborn.

[20] Pfeifer, R. and Scheier, C. (2001), *Understanding Intelligence*, The MIT Press: Cambridge, Mass.

[21] Newell, A. and Simon, H. A. (1972), *Human Problem Solving*, Englewood Cliffs NJ.

[22] Valera, F. J., Thompson, E., and Rosch, E. (1991), *The Embodied Mind. Cognitive Science and Human Experience*, The MIT Press: Cambridge, Mass.

[23] Marcus, G. (2003), *The Algebraic Mind: Integrating Connectionism and Cognitive Science*, The MIT Press: Cambridge, Mass.

[24] Mainzer, K. (2009), From embodied mind to embodied robotics: Humanities and system theoretical aspects, *Journal of Physiology* (Paris), 103, pp. 296–304.

[25] Wilson, E. O. (2000), *Sociobiology. The New Synthesis*, 25th Anniversary edn., Cambridge, Mass.

[26] Mataric, M. J., Sukhatme, G. S., and Ostergaard, E. H. (2003), Multi-robot task allocation in uncertain environments, *Autonomous Robots,* 14(2–3), pp. 253–261.

[27] Brooks, R. A. (2005), *Menschmaschinen*, Frankfurt.

[28] Lee, E. A. (2008), Cyber-physical systems: Design challenges, University of California Berkeley Technical Report No. UCB/EECS-2008-8.

[29] acatech (Ed.) (2011), *Cyber-Physical Systems. Innovationsmotor für Mobilität, Gesundheit, Energie und Produktion*, Berlin.

[30] Tretmans, J. and Brinksma, E. (2003), TorX: Automated model-based testing, in Hartman, A. and Dussa-Zieger, K. (Eds.) *Proceedings of the First European Conference on Model-Driven Software Engineering*.

[31] Mainzer, K., Schuster, P., and Schwichtenberg, H. (Eds.) (2018), *Proof and Computation. Digitalization in Mathematics, Computer Science, and Philosophy*, World Scientific: Singapore.

[32] Coquand, T. and Huet, G. (1988), The calculus of constructions, *Information and Computation*, 76(2–3), S. 95–120.

[33] Coupet-Grimal, S. and Jakubiec, L. (1996), Coq and hardware verification: A case study, in *TPHOLs*, LCNS, Vol. 1125, pp. 125–139.

[34] Mainzer, K., Schuster, P., and Schwichtenberg, H. (Eds.) (2021), *Proof and Computation. From Proof Theory and Univalent Mathematics to Program Extraction and Verification*, World Scientific: Singapore.

[35] Kant, I. (1900 ff.), Ausgabe der Preußischen Akademie der Wissenschaften. Berlin, AA IV, 421: *Handle nur nach derjenigen Maxime, durch die du zugleich wollen kannst, dass sie ein allgemeines Gesetz werde.*

[36] Mainzer, K. and Kahle, R. (2022), *Limits of AI — Theoretical, Practical, Ethical,* Springer: Berlin.

[37] Ouyang, L. *et al.* (2022), Training language models to follow instructions with human feedback, *arXiv:* 2203,02155vl [cs.CL], 4 March 2022.

[38] Mainzer, K. (2023), ChatGPT and artificial intelligence: From foundations to applications in education, *Peking University Education Review Journal,* 1 (Chinese).

[39] Gogoll, J., Heckmann, D., and Pretscher, A. (2023), Endlich neue Prüfungen dank ChatGPT, FAZ 20.3.2023 Nr. 67, 18.

[40] Mainzer, K. (2023), ChatGPT and artificial intelligence: From foundations to applications in education, *Peking University Education Review Journal,* 1 (Chinese).

[41] Wenfeng Liang *et al.* (2025), DeepSeek-R1: Incentivizing reasoning capability in LLMs via reinforcement learning, *arXiv:* 2501.12948 [cs.CL].

[42] Pengfei Liu, Weizhe Yuan, Jinlan Fu, Zhengbao Jiang, Hiroaki Hayashi, and Graham Neubig (2021), Pre-train, prompt, and predict: A systematic survey of prompting methods in natural language processing, *arXiv:* 2107.13586v1 [cs.CL], 28 July 2021.

[43] Montti, R. (2022), Google's chain of thought prompting can boost today's Best4Algorithms, *Search Engine Journal,* 13 May 2022.

[44] Vidrih, M. (2025), How does DeepSeek-AI model work — Simplified. https://vidrihmarko.medium.com/the-deepseek-r1-model-simplified-8fde88abbe07.

[45] Zuckerman, G. (2019), *The Man Who Solved the Market: How Jim Simons Launched the Quant Revolution,* Portfolio Penguin: New York.

[46] Aaronson, S. (2019), Why Google's quantum supremacy milestone matters, *The New York Times,* 30 October 2019. https://www.nytimes.com/2019/10/30/opinion/google-quantum-computer-sycamore.html (abgerufen 10 May 2020).

[47] IBM Quantum Update: Q System One Launch, New Collaborators, and QC Center Plans. https://www.hpcwire.com/2019/01/10/ibm-quantum-update-q-system-one-launch-new-collaborators-and-qc-center-plans/ (abgerufen 10 May 2020).

[48] Mainzer, K. (2030), *Quantencomputer. Von der Quantenwelt zur Künstlichen Intelligenz,* Springer: Berlin.

[49] Audretsch, J. and Mainzer, K. (Eds.) (1996), *Wieviele Leben hat Schrödingers Katze? Zur Physik und Philosophie der Quantenmechanik,* Spektrum Akademischer Verlag: Heidelberg.

[50] Preskill, J. (2018), Quantum Computing in the NISQ era and beyond, *arXiv:* 1801.00862v3 [quant-ph].

[51] Aharonov, D., van Dam, W., Kempe, J., Landau, Z., Lloyd, S., and Regev, O. (2007), Adiabatic quantum computation is equivalent to standard quantum computation, 37, pp. 166–194.

[52] Shor, P. W. (1997), Polynomial-time algorithms for prime factorization and discrete logarithms on a quantum computer, *SIAM Journal on Computing*, 26(1997), pp. 1484–1509.

[53] Hidary, J. D. (2019), *Quantum Computing: An Applied Approach*, Springer: Cham, pp. 65–66.

[54] Briegel, H.-J., Dür, W., Cirac, J. J., and Zoller, P. (1998), Quantum repeaters: The role of imperfect local operations in quantum communication, *Physical Review Letters*, 81, S. 5932–5935.

[55] Enabling technology, *Business Dictionary*. http://www.businessdictionary.com/definition/enabling-technology.html#ixzz2lrYdBsg3 (abgerufen 12 May 2020).

[56] Mainzer, K. (2014), *Die Berechnung der Welt. Von der Weltformel zu Big Data*, C. H. Beck: München.

[57] Mainzer, K. (2018), *Wie berechenbar ist unsere Welt? Herausforderungen für die Mathematik, Informatik und Philosophie im Zeitalter der Digitalisierung*, Springer Essentials: Wiesbaden.

[58] Europäische Kommission (2017), The Impact of Quantum Technologies on the EU's Future Policies. Part 1 Quantum Time. Luxemburg.

[59] Kagermann, H., Süssenguth, F., Körner, J., and Liepold, A. (2020), *Innovationspotentiale der Quantentechnologien der zweiten Generation*, acatech Impuls, Chapter 8.2.

[60] BMBF (Ed.) (2018), *Quantentechnologien — von den Grundlagen zum Markt*, Rahmenprogramm der Bundesregierung: Bonn.

[61] Jones, R. A. L. (Ed.) (2018), Quantum Technologies. Public Dialogue Report, EPSRC, UK Research and Innovation.

[62] Mainzer, K. and Chua, L. (2013), *Local Activity Principle*, Imperial University Press: London.

[63] Ivanov, P. C. (2021), The new field of network physiology: Building the human physiolome, *Frontiers in Network Physiology*, 1:711778.

[64] Costa, F. P., Tuszynski, J., Iemma, A. F., Trevizan, W. A., Wiedenmann, B., and Schöll, E. (2025), External low energy electromagnetic fields affect heart dynamics: Surrogate for system synchronization, chaos control and cancer patient's health, *Frontiers in Network Physiology*, 4, p. 1525135.

[65] Banerjee, R. and Chakrabarti, B. K. (2008), *Models of Brain and Mind. Physical, Computational, and Psychological Approaches*, Progress in Brain Research: Amsterdam.

[66] Chua, L. (1971), Memristor: The missing circuit element, *IEEE Transactions on Circuit Theory* 18(5), pp. 507–519.

[67] Chua, L. (2014), If it's pinched it's a memristor, *Semiconductor Science and Technology*, 29(10), pp. 104001–1040042.

[68] Sah, M. P., Kim, H., and Chua, L. O. (2014), Brains are made of memristors, *IEEE Circuits and Systems Magazine*, 14(1), pp. 12–36.

[69] Stanley Williams, R. (2008), How we found the missing memristor, *IEEE Spectrum*, 45(12), pp. 28–35.

[70] Tetzlaff, R. (Ed.), *Memristors and Memristive Systems*, Springer: Berlin, p. 14 (similar to Fig. 1.5).

[71] NeuroSys (2021), BMBF Future Cluster Neuromorphic Hardware for autonomous Systems of Artificial Intelligence, Projekt Paper, pp. 14–15.

[72] Mirsa, J. and Saha, J. (2010), Artificial networks in hardware: A survey of two decades of progress, *Neurocomputing*, 74(1–3), pp. 239–255.

[73] Shasin, B. J., Tait, A. N., de Lima, T. F., Pernice, W. H. P., Bhaskaran, H., Wright, C. D., and Prucnal, P. R. (2020), Photonics for artificial intelligence and neuromorphic computing. *arXiv*: 2011.001111.

[74] Kickuth, R. (2021), *Bio-inspired Computing. Gehirn-künstliche Neuronetze — neuromorphe Architekturen ... und wie es weitergeht: Photonik-Quantencomputer-Evolution.* Rubikon.

[75] Mainzer, K. and Kahle, R. (2022), *Limits of AI — Theoretical-Practical-Ethical*, Springer: Berlin.

[76] Ulmann, B. (2010), *Analogrechner: Wunderwerke der Technik — Grundlagen, Geschichte und Anwendung*, Oldenbourg: München.

[77] Mainzer, K. (2018), *The Digital and Real World. Computational Foundations of Mathematics, Science, Technology, and Philosophy*, World Scientific: Singapore, Chapters 10–12.

[78] Ulmann, B. (2010), *Analog and Hybrid Computer Programming*, De Gruyter: Berlin.

[79] Mainzer, K. (2020), *Leben als Maschinen*, 2nd edn., Brill Mentis: Paderborn.

[80] Mainzer, K. (1997), *Gehirn-Computer Komplexität*, Springer: Berlin.

[81] Hiesinger, P. R. (2021), *The Self-Assembling Brain. How Neural Networks Grow Smarter*, Princeton University Press: Princeton, NJ.

[82] Wahlster, W. and Winterhalter, C. (Eds.) (2020), *German Standardisation Roadmap Artificial Intelligence*, DIN + DKE: Berlin.

[83] Intel Technology Brief Loihi 2: Taking neuromorphic computing to the next level with Loihi 2, © Intel cooperation 0921/SBAI/KC/PDF. Intel Cooperation 2021: Lava Software Framework (https://lava-nc.org/).

[84] Müller, E. *et al.* (2022), The operating system of the neuromorphic BrainScaleS-1 system, *arXiv: 2003.13749v2* [cs.NE] 2 Feb 2022.

[85] Delle Donne, C. *et al.* (2025), An operating system for executing applications on quantum nodes. *Nature*, 639, pp. 321–328.

Chapter 5

Competition of Innovative Systems

A key message of this book is that future technologies are bets on the future. Together with current initial and bridging technologies, they must be combined in a sustainable innovation portfolio in order to meet the challenges of the future. Only a good innovation portfolio is in a position to face crises and the constantly changing diversity and complexity and to adapt to them in a resilient manner in order to at least make a 'good cut' in the end.

Another important insight is that innovation is by no means limited to technology and machines. Despite all the AI and machine learning, human creativity is an essential component of innovation clusters. This is why we also talk about hybrid innovation clusters in which different technologies are combined. Above all, people provide the normative orientation for innovations.

With a view to humanity as a whole and the future of this planet, they should be geared towards sustainability goals. In reality, however, there is a competition between different political value systems that is being fought out with innovations. In peaceful and fair disputes, the competition is primarily played out on the markets. In the language of complex dynamic systems, stable equilibria can be found, in which the advantages and disadvantages of action strategies can be analysed. However, war and terror are still among the options of non-linear dynamics that make the entire Earth system chaotic and uncontrollable.

Expenditure on research and development

One measure of a country's innovation is the innovation survey. Take a medium-sized European country such as Germany as an example. In its 2021 Innovation Survey, the Centre for European Economic Research (ZEW) revealed that innovation expenditure by companies in Germany fell for the first time in 10 years in 2020 [1]. After all, this represents a decrease of 3.6% to 170.5 billion euros. This was due to the slump caused by the coronavirus crisis. Such slumps are serious, as innovations in products, business models and production processes — as Schumpeter already emphasised — are key factors for a society's economic growth and prosperity. The resilience of the economic system, i.e. the ability to recover after a slump, will depend crucially on whether investments in innovation and thus the innovation dynamic can be resumed.

The first important factor in innovation investment is spending on research and development. Before the coronavirus slump in 2019, the private and public sectors invested a combined €110 billion in Germany. This amount corresponds to 3.2% of gross domestic product (GDP). In this case, Germany even exceeded the EU target of 3%. Germany was therefore on the way to achieving its self-imposed target of 3.5% of GDP in 2025. This development should be recognised in light of the previous trend, according to which expenditure still amounted to 2.9% in 2014. Measured in terms of GDP, investment in research and development in 2018 was among the highest in the world:

South Korea	5.5%
Switzerland	3.4%
Sweden	3.3%
Japan	3.3%
Germany	3.2%
USA	2.8%
France	2.2%
Netherlands	2.2%
China	2.1%
UK	1.7%

Investment does not equal innovation. Only at first glance does the balance sheet for Germany look positive, but in an international comparison of innovation performance, it is not good enough. In global innovation rankings, Germany did not occupy any top positions on average, which does not mean that they were achieved in some specialised sectors. According to the Global Innovation Index of the World Intellectual Property Organisation (WIPO) of the United Nations (UN), Germany's innovative strength was ranked 10th in 2021. The frontrunners were Switzerland, Sweden and the USA:

Switzerland	65.5 points
Sweden	63.2 points
USA	61.3 points
UK	59.8 points
South Korea	59.3 points
Netherlands	58.6 points
Finland	58.4 points
Singapore	57.8 points
Denmark	57.3 points
Germany	57.3 points
France	55.0 points
China	54.8 points
Japan	54.5 points

Innovation power with patents

However, such rankings are not set in stone, but depend on the respective weightings and assessment criteria. An internationally recognised measure of a country's innovative strength refers to the number of patents registered by the country internationally. Patents are also associated with industrial property rights for companies' products, which mark their lead over competitors in their own country and internationally. WIPO records patent applications in an international patent system [2]. China now tops the list of new applications worldwide, ahead of the USA, with Germany in fifth place. However, a distinction must be made between the quality of patents, which varies from country to country.

In terms of disciplines, most patent applications are in computer technology, digital communication, medicine and electrical engineering. In terms of companies, the Chinese IT group Huawei is in the lead, followed by the US semiconductor manufacturer Qualcomm and the South Korean IT groups, Samsung and LG. Germany is traditionally strong in innovation in the automotive industry with corresponding patents. Bosch and Siemens registered the most international patents in Germany.

Innovative strength depends crucially on the educational institutions in the individual countries. The level of education creates knowledge and expertise as a prerequisite for the development of patents. Four universities in the USA and four universities in China lead the world's educational institutions with the most patent applications. In continental Europe, Switzerland has a particularly high proportion of academics specialising in technology and natural sciences. North America and the UK also invest particularly heavily in this area of education.

Knowledge transfer from science to industry

As already emphasised several times, the transfer of knowledge from science to industry is fundamental to successful innovation. Chapter 3 highlighted the academic spin-offs that turn inventions from universities into start-ups. Germany in particular struggles in this area with its cumbersome bureaucracy and its legal jungle of paragraphs. In addition, the necessary funds are easier to access in the USA, for example, than in Germany, which is also linked to the respective financial and tax systems. The Technical University of Munich (TUM) is an exception by German standards. But, as we all know, one swallow does not make a summer.

Chip industry as the game changer of innovation power

Chips are considered to be the most important building block of all current and future innovation boosts. The global turnover of the semiconductor industry can be measured in investments. According to forecasts by the industry association SEMI [3], around 400 billion dollars will be invested in new systems, factories and plants by 2027. More than 100 billion dollars has already been achieved. In the three years after 2025, China will lead with 100 billion dollars, followed by Korea with 81 billion dollars, Taiwan with 75 billion dollars and the USA with 63 billion dollars. They generally invest 20% of their profits in their own capitalisation, research and development.

The drivers of this development are national corporations [4]. The largest chip companies by market capitalisation in billions of dollars are as follows:

Nvidia (USA)	3335
Broadcom (USA)	1044
TSMC (Taiwan)	1044
ASML (Netherlands)	286
Samsung (South Korea)	244
AMD (USA)	192
Qualcomm (USA)	172
Texas Instrument (USA)	170
Arm Holdings (UK)	148
Applied Material (USA)	140

While Europe is falling behind, China and the USA are engaged in a systems competition. The USA feels threatened in its global supremacy, while American chip sanctions are perceived by China as an obstacle to competition. Monopoly commissions in both countries control takeovers worth billions. Pre-products for the chips are also subject to strict scrutiny.

China is now the largest manufacturer of silicon wafers, on which electronic semiconductor components are produced. The following number of chip wafers were produced in millions per month in 2024:

China	8.6
Taiwan	5.7
South Korea	5.1
Japan	4.7
USA	3.1
Rest of the world	4.4

Concerning contract manufacturers, TSMC dominates the chip market with market shares in %:

TSMC	61.7
Samsung Foundry	11.0
UMC	5.7
SMIC	5.7
GlobalFoundries	5.1
HuaHong Group	2.2
Tower	1.1
PSMC	1.0
Nexchip	1.0
VIS	1.0

Europe, on the other hand, is clearly falling behind. The relegated US chip manufacturers Intel and Wolfspeed withdrew their investment plans for Germany, Poland and France totalling 35 billion euros. Since 2023, the EU Chip Act should open up investment subsidies of 42 billion euros by 2030 in order to anchor Europe more firmly in the global chip supply chains and more than double the global market share of semiconductor components manufactured in the EU to 20%. According to current estimates, these plans will not be sustainable either. We will have to wait and see how TSMC's investment package of 10 billion euros for a joint plant in Dresden, Germany, develops.

Almost half of all modern chip factories are located in China and the USA. Korea and Taiwan have also announced subsidies in the three-digit billion dollar range. After limiting itself to the construction of machines for the manufacture of chips in the 1980s, Japan is returning to chip production. The number of chip factories in 2023 was as follows:

North America	342
China	303
Japan	248
Taiwan	172
South Korea	87
Asia Pacific	63
Europe	60
Rest of the world	195

The type of chips that are produced is crucial for the future strategy. For too long, Europe has only focussed on simple industrial chips that are used in cars and trains, coffee machines and washing machines, for example. For the future with smartphones and AI, however, smaller and finer structures are needed. TSMC is a leader in this area, serving customers, such as Nvidia and Apple.

A competitiveness masterplan for Europe?

Artificial intelligence is not only about intelligent algorithms but also about training data. Metaphorically speaking, data is the oil of the AI age. This is also a new opportunity for countries like Germany. Germany was traditionally strong in areas such as industrial and plant engineering, mechanical engineering, the automotive industry, chemicals and pharmaceuticals, and, as an export nation, Germany was in the Champions League in the 20th century. However, the 21st century is and will be the century of artificial intelligence. Countries like Germany have a huge wealth of data from the experience of the traditional industrial world, which needs to be utilised with modern AI. This is also the competitive advantage over the new AI nations. Europe just needs to learn how to harness this treasure trove of data and not hinder it through national parcelling out and regulation.

Incidentally, this also applies to one of the largest countries such as Russia, which is no longer even mentioned in the battle between the world's leading IT and AI nations. Russia leads the way with its vast natural resources from gas and oil to uranium, for example. In the current transition phase, such natural advantages undoubtedly play an important role. But they are not the key to the future. A smart policy would utilise the vast amounts of data on the experience with these natural resources to build the AI-supported supply systems and infrastructures of the future.

A key obstacle in Europe (especially Germany) is the tendency to over-regulate for perfection. One example is the EU's AI Act, which stifles innovation dynamics in start-ups and medium-sized companies with a complicated regulatory framework. The ugly face of regulation and ethical rules is a bureaucracy that threatens to stifle any creative approach with its complicated and sometimes inconsistent requirements. While the USA is coming up with an AI funding programme worth billions called 'Stargate' and China is responding with outstanding start-ups such as DeepSeek, Europe is reacting with a tightly meshed regulatory framework, over 80% of which only talks about the dangers and risks of AI. Germany must also transpose this European AI law into national law, which could end up placing an additional burden on Germany as a centre of innovation.

Instead, the motto should be 'Invest instead of regulate'! And investing does not just mean spending trillions of dollars, but finding intelligent, effective and energy-saving solutions. The Chinese–American competition shows that scaling with even more money and even more energy is not expedient and is reaching its limits in the sense of the limit value theory. Breakthroughs always come in the end through investments in the natural resource of mind!

In the second term of office of the current President of the European Commission from 2025, the focus will be on the competitiveness of the EU instead of the Green Deal. The focus will be on digitalisation and the energy sector in order to make the Green Deal a success. Modelled on the US innovation agency Advanced Research Project Agencies (ARPA), an agency will be set up to drive investment in key technologies such as AI. In a special report, the former President of the European Central Bank Mario Draghi warned that Europe was at risk of falling into decline if the lack of digitalisation in Europe was not remedied. Draghi put the investment requirement at 750–800 billion EUR per year. The EU is now talking about the creation of 'mega-factories' for artificial intelligence that could be used by research and industry for training. Following the example of the Geneva Nuclear Research Centre, the aim is to create a 'CERN for AI' in order to use AI primarily in key sectors of industry such as the automotive sector.

There is a big question mark over whether it will be possible to bundle industrial policy in Brussels. Larger European countries such as France and Germany pursue a traditionally strong but nationally determined industrial policy. In global competition with major powers such as the USA and China, this can impair competitiveness and efficiency. From the point of view of complex systems, it is important to create an efficient competitiveness coordination instrument. The aim is to coordinate the various European states and align them with jointly selected priorities. If this could be achieved for the energy and digital infrastructure together with AI, it would be a key step forward for Europe. Such a measure would have to be supported by a competition fund.

In future, competition controls should serve to promote innovation. Bureaucracy should also be reduced by eliminating or bundling reporting obligations (e.g. Supply Chain Act). A capital markets union should facilitate access to venture capital. The following overview of the development (in billions of dollars) in the period from 2015 to 2024 shows the extent to which global AI start-up financing depends on venture capital [5]:

2015	10
2016	20
2017	29
2018	37
2019	38
2020	40
2021	83
2022	54
2023	56
2024	101

Simpler rules in insolvency, labour and tax law are intended to support start-ups. This is intended to keep companies in Europe that would otherwise emigrate to the USA, for example. In order to reduce the comparatively high energy prices relative to China and the USA, a suitable innovation portfolio must be determined.

The innovation push behind DeepSeek

The study of complex dynamic systems shows that states and organisations can only provide the financial and economic framework conditions to help innovations achieve a breakthrough. Innovations are the result of local activity, just like mutations in biological evolution. As explained in Chapter 2, local activity destabilises existing equilibria of a complex system (societies, markets) by creating new orders and structures. Initially, minimal energy reinforces itself in order to trigger a 'breakthrough' at a certain threshold value. This can be a neuron in the brain, whose state of tension is self-reinforced by adding up small impulses until the cell finally 'fires' at a certain threshold value. In human societies, it is individual agents that add up their experiences and finally come up with a new realisation at a certain threshold value. Joseph Schumpeter called such economic agents 'entrepreneurs'.

The innovation of DeepSeek with its developer Liang Wenfeng is a classic example of this. As already mentioned, Liang Wenfeng was interested in maths at school, but first studied engineering before founding the investment fund High-Flyer-Capital with two colleagues. The motivation behind this is not just to make money. In recent years, Liang has repeatedly referred to the American mathematician and hedge fund manager James Harris Simons (1938–2024), who apparently fascinated him [6].

Simons was originally a theoretical mathematician who worked on minimal surfaces in topology and differential geometry, which are used in string theory, knot theory and topological quantum field theory. He received high honours from the American Mathematical Society for his theoretical results in foundational mathematical research. In 1978, he switched to the financial sector, where he applied advanced mathematical methods for the first time. Simons developed a system with which developments in the financial markets can be mathematically modelled and predicted.

This marked the start of a breathtaking career from a small investor to one of the most successful hedge fund managers in the world. This career began in 1982 in New York with the founding of the investment company Renaissance Technologies (RenTech). In 2006, Simons was named 'Financial Engineer of the Year' by the International Association of Financial Engineers (IAFE). In 2020, his private wealth totalled 23.5 billion US dollars. This made him the 23rd richest American and the 36th richest person in the world. In 2019, he was listed as the highest-earning hedge fund manager.

Nevertheless, Simons remained true to his original love of theoretical mathematics. He donated 60 million dollars to the Science Centre for Geometry and Physics at Stony Brook. His Simons Foundation promotes research in mathematics and theoretical physics. His interest in the foundations of mathematical sciences is also reflected in the fact that he was a member of the American Philosophical Society from 2007 and a member of the American Academy of Arts and Sciences from 2008. In 2012, he became an honorary member of the London Mathematical Society.

According to the Bloomberg financial agency, RenTech has earned 55 billion dollars from the computer codes of its mathematical models over the decades since it was founded, with an average return of 40% after fees. It is no wonder that this fund (also known as 'Quant') is still highly attractive today to intelligent users, such as Liang Wenfeng from Hanghzou. This Chinese metropolis is a high-tech centre where Chinese high finance and digital companies such as Alibaba are concentrated.

Nevertheless, Wenfeng appears to have pursued an innovation strategy independent of big money. He is not only inspired by Simon's life path. By 2021, his hedge fund High-Flyer will have systematically implemented many of RenTech's strategies. AI is already helping to extract key data from huge data sets that can be useful for predicting share prices and making investment decisions. The next step was to focus entirely on the fundamentals of the AI used with a new start-up, regardless of its application in the financial world. This is DeepSeek's innovation strategy. The final vision on this path is a Artificial General Intelligence (AGI) that is superior to human intelligence from a certain critical phase transition onwards. This critical point is called the 'singularity'.

DeepSeek Innovation is therefore interested in the basics — independent of big commerce. This does not exclude the possibility of accepting the big profit afterwards as a bonus. This strategy is encouraging for all those intelligent people around the world, who have already given up in the face of the far superior concentration of capital in the USA in the competition for AI, for example, in Europe.

A major disadvantage for European AI companies is the lack of access to computing power and the energy required for this. When an American president holds out the prospect of not only trillion-dollar funding programmes but also his own nuclear power plants to provide the necessary energy for such huge digital data centres, the rest of the world is left with nothing but resignation.

> However, the important lesson from DeepSeek is that naive scaling in terms of money, energy, manpower and computing power does not solve the increasingly difficult problems. This was demonstrated a few years ago in the EU's Brain Project, when it was believed that the secrets of the brain could be unravelled by not only investing millions of euros but also by concentrating all the leading computer, neuroscientists and cognitive scientists in one mega-project. This project fell apart in a dispute over expertise. It was criticized that there was no common key idea between biological brain research and computer science [7]: Neurobiologists want to understand the biological brain. Computer scientists are only inspired by the brain in order to solve computational problems with effective technology.

DeepSeek was created under the pressure of restrictive framework conditions. Due to the American boycott, the Chinese chip market had no access to the latest cutting-edge chip technology, which seemed unavoidable for the further development of AI. They had to make do with older technology and less energy. It therefore came down to an intelligent solution as to how to find at least a comparatively good, if not better, solution under these conditions.

From a European perspective, it is noteworthy that DeepSeek uses an architecture known from the French start-up Mistral [8].

The system is called 'Mixture-of-Experts': Instead of using the entire AI program for each task, a router forwards the given task to a subroutine that specialises in this problem. In other words, the AI consists of a group of highly specialised experts for different tasks. This expert then only accesses its specialised sub-data and not the huge reservoir of all the data of the entire AI. This means that considerably less energy, computing power and computing time are used overall. It's like a professional human problem solver. Only the beginner sits in front of a task with his head spinning and overstrains his brain in the search for a solution by activating all his knowledge and skills. The cool professional recognises the special area of this task and the tools required for it. Only this sub-area is then activated, saving resources, effort, time and energy.

In the DeepSeek research paper from December 2024, a team of almost 200 authors reports on how the DeepSeek language model was developed. 'Mixture-in-Experts' is just one important principle among many other details, most of which are known. It depends on the skilful combination of all these details, which could only be achieved through engineering teamwork. This software had to be adapted to 2048 H800 Nvidia chips, which are by no means among the highest quality Nvidia chips, but were the only type available to the Chinese developers. However, as has been emphasised several times, algorithms, software and hardware are not the only characteristics of a successful AI system. The training data is also crucial. But the Deep Seek team provides just as little information about this as the competition. Mutual suspicions among competitors that they have violated copyrights when training their data sets are commonplace.

What remains of DeepSeek at the time of writing (February 2025) is that this AI is significantly more efficient and uses less energy. This message has also reached American competitors, who have expressed their appreciation of the Chinese competitor. In this way, a cloud provider such as Microsoft could offer more customers even more AI enhanced with augmented reality, for example. Facebook points out that DeepSeek confirms the Meta strategy with open source for all users. Like DeepSeek, Meta is making its AI code (Llama) public so that others can build on it. The advantage of the open source strategy is that the ideas, results and methods of other users can be incorporated into one's own strategy.

> What should be emphasised once again, however, is DeepSeek's focus on long-term basic research with a (philosophical) vision and not just on short-term commercial successes that are to be achieved with a 'tonne ideology' (scaling): Even more money and even more energy will not win the race in the end. No, it is the 'raw material spirit' that we need to invest in!

References

[1] Centre for European Economic Research (ZEW). https://www.zew.de/publikationen/zew-gutachten-und-forschungsberichte/forschungsberichte/innovationen/innovationserhebung.

[2] World Intellectual Property Organization (WIPO). https://www.wipo.int/portal/en/index.html.

[3] Semiconductor Association. https://www.semi.org/eu.

[4] Statista data research (S. Finsterbusch). *Sources*: Boston Consulting Group, SIA, CSIS, Global Data, Statista Market Insights.

[5] Precedence Research; CAICT; KPMG; Statista; Crunchbase.

[6] Zuckerman, G. (2019), *The Man Who Solved the Market: How Jim Simons Launched the Quant Revolution*, Portfolio Penguin: New York.

[7] Frégnac, Y. and Laurent, G. (2014), Neuroscience: Where is the brain in the Human Brain Project? Europe's €1-billion science and technology project needs to clarify its goals and establish transparent governance, *Nature*, 513 (7516), pp. 27–29.

[8] https://mistral.ai/products/le-chat.

Chapter 6

Future between Sustainability and Planetary Exploration

On this planet, we are developing into a long-term state of global internal politics for which there is no longer any outside. The global challenges facing humanity, such as pandemics, climate and world food crises, as set out in the UN's Sustainable Development Goals, for example, are actually forcing us to act together quickly. Despite all the technical and innovative progress, however, humanity is proving to be shockingly resistant to learning in the social and political sphere and is repeatedly reverting to the archaic behavioural patterns of its species. Wars of unprecedented cruelty are justified, as they were a 100 years ago, with ideologies that were thought to be outdated.

In Europe, we are once again seeing images of armoured battles like those that raged during the Second World War. Those responsible for these battles only cause immeasurable suffering and misery to the civilian population, destruction and annihilation of civilian infrastructure, without solving a single problem. These completely pointless and superfluous learning loops, which humanity repeatedly allows itself, not only destroy the basis of life in the respective countries, but, in the case of technically and economically highly developed countries, they also destroy global supply chains and trigger global supply crises among the poorest of the poor on this planet far away from the theatre of war. These are typical nonlinear effects of the complex societal dynamics.

These senseless wars destroy resources and waste time that is urgently needed to combat global threats to this planet. The fact that climate collapse on this planet is inevitable if immediate action is not taken in line with the known critical control values is completely pushed into the background. Because people are so distracted with their wars and their party-political disputes with each other, the red line with nature could end up being crossed and the planet could slide uncorrectably into global catastrophe.

But it's not just the planet's ecological problems. Occasionally, science fiction films drastically demonstrate that this planet is just a tiny, vulnerable celestial body in a completely hostile environment. The universe is not an eternal event of harmony, as previous centuries had imagined. The 1998 film *Deep Impact* drastically demonstrated a scenario that could be statistically quite probable at present: If a comet hurtles towards the Earth and threatens total annihilation, then there is nothing left to do but forget all conflicts among ourselves, co-operate worldwide and, in the short term, pool all of humanity's innovative potential to solve this ultimate problem. If it were not so terrible, one would occasionally wish for such a 'heavenly' hint.

Universe as innovation space

However, space is not only a threat but also a fascinating opportunity for mankind. When the author began his studies in mathematics, physics and philosophy in 1968, Stanley Kubrick's masterpiece *2001: A Space Odyssey* was released in cinemas. This film is still moving today. It is about the evolution of man, his symbiosis with artificial intelligence and his departure into space. The existential questions posed by the film, its philosophical mood staged with economical means and not least its technically and physically correct portrayal of space travel and robotics (by the standards of the time) are still impressive today.

After the moon landing in 1969, it seemed as if landing on Mars, Jupiter and other celestial bodies was only a matter of a few years away. The two superpowers at the time, the USA and the USSR, were in fierce competition. But after the dissolution of the Warsaw Pact, they finally recognised a common challenge in the issue of space travel and created joint research and development formats in order to pool resources for this enormous innovation task for mankind. Subsequently, these joint projects were jeopardised by conflicts between the superpowers. However, the superpowers have resumed their space projects as nation states. The United States is now announcing the journey to Mars as its next goal. What is remarkable and new is that large tech companies are now taking on space programmes alongside nation states. Similar to AI research, research and development is shifting to innovative companies.

Artificial Intelligence and robotics as space innovation

Artificial intelligence (AI) and robotics are the technical realisation of algorithms, without which space travel would not be possible. Whether biological humans as we know them today will even be able to develop cosmic civilisations is an open question. In any case, AI-based technology is already travelling in space with space probes and robots. This is not just about expanding knowledge and insight but also about energy issues and political–military power. AI algorithms make all of this possible. But will they also be able to take on a life of their own, as in Kubrick's film, either to overcome the inability of a divided humanity or to develop their own interests?

Occasionally, nature is the model for technical developments. However, computer science and engineering often find solutions that are different and even better and more efficient than those found in nature. Aircraft construction and ultimately space travel were only successful when people stopped building flying machines modelled on birds. Only the utilisation of the laws of aerodynamics and physics led to technical solutions that far surpassed known evolutionary developments. There is therefore no such thing as 'the' intelligence, but rather degrees of efficient and automated problem solving that can be realised by technical or natural systems.

Natural intelligence arose through evolution. It therefore makes sense to simulate evolution using algorithms. Genetic and evolutionary algorithms are now also used in technology. Life science and computer science are merging in neuromorphic computer structures. Biological brains not only enable amazing cognitive performances such as seeing, speaking, hearing, feeling and thinking, they also work much more efficiently than energy-guzzling digital supercomputers.

Combined with supercomputers, neural networks and learning algorithms now surpass the learning ability of humans when it comes to recognising patterns. Today, we call this machine learning. AI today means machine learning — a big hype in technology and business. However, neural networks with thousands of parameters turn out to be 'black boxes' whose causal processes between the individual neurons are not transparent. Ultimately, it is just statistical learning based on data correlations, in which neural networks are trained like the brains of small children or animals with a finite number of data sets. Greater explainability of neural networks is a central challenge of research. Questions of liability and responsibility can only be clarified if there is a clear causal distinction between causes and effects. Whether and how causal learning can be mapped in algorithms is currently the subject of research. This would be the first step from weak (statistical) AI to strong AI.

Machine intelligence as weak AI is already becoming the basis of an increasingly automated working world, which is already being realised on Mars. On Earth, the development is moving from machine assistance systems to humanoid robots with a human-like appearance that interact with people in the workplace and in everyday life. In a stationary industrial robot, work steps are defined in a computer programme. Social and cognitive robots, on the other hand, must learn to perceive their environment, make decisions and act independently. Intelligent software must be combined with sensor technology in order to realise this type of social intelligence (embodied robotics). Even in an organism, brains are not enough; bodily sensations must be taken into account to experience the world (embodied mind). This will be crucial for use on distant planets.

Traditionally, we associate intelligence with a single organism such as a human person. Kubrick's film already correctly showed that the intelligent on-board computer HALE does not have to be connected to a human-like organism. Behind this is often just the vanity of people, who already projected their intelligence into human-like images of gods in their religions and could only imagine artificial intelligence in human-like robots à la Golem.

In fact, intelligence has long since been realised to a certain extent in anonymous global networks, such as the Internet, which are increasingly equipped with self-learning algorithms. As a result, a new concept of intelligence is emerging that is invisibly distributed in infrastructures. Global mobility networks are one example. Cars are already being described as computers on four wheels. As autonomous vehicles, they generate intelligent behaviour that more or less completely replaces the human driver. What application scenarios are associated with this in transport systems? As swarm intelligence shows, intelligence is by no means limited to individual organisms, even in nature. In the Internet of Things, objects and devices can be equipped with intelligent software interfaces and sensors to solve problems collectively. Another current example is the Industrial Internet (Internet 4.0), in which production and sales organise themselves largely autonomously.

According to our working definition, a factory is called intelligent if it can perform complex tasks more or less independently. We now generally speak of cyber–physical systems, which also include smart cities and smart grids. Cosmic space stations on distant planets will be inconceivable without these intelligent infrastructures.

Since its inception, AI research has been associated with grand visions of the future of humanity. Will there be neuromorphic computers that can not only simulate the human brain but also replace it in order to survive in a hostile space environment? How do analogue processes of nature differ from digital technology? Are the technologies of artificial life from systems biology and synthetic biology converging with artificial intelligence? Will there be a general artificial intelligence that is superior to humans?

From hybrid AI to superintelligence?

So far, it has been shown how AI systems outperform humans in certain areas. AI systems can calculate, answer and ask questions faster, recognise and predict correlations, handle larger amounts of data, have a more extensive memory, etc. But could there also be a hybrid AI system that is superior to humanity's collective intelligence? Then we are talking about superintelligence [1]. The following criteria would have to be considered:

(1) **Fast superintelligence:** An AI system can do all that humans can do, only faster.

We know the experience of being faster than others or others being faster than us. In the first case, we become increasingly bored. In the other case, we feel overwhelmed. Increases in the performance of AI systems based on silicon or organic materials such as carbon in nanoelectronics according to Moore's law are conceivable. However, neuromorphic systems using neuronal tissue developed in synthetic biology would also be possible. This tissue could be implanted in living brains both in the event of damage and for performance enhancement. Enhancement of the natural brain is ethically controversial and would also come up against medical–biological limits. The technical possibilities of quantum computers would surpass everything that has gone before in terms of speed, but would not change the logical–mathematical limits and laws of computability.

(2) **Collective superintelligence:** An AI-system consists of many partial systems which can realise less than humans. But the collective system surpasses individual humans.

This strategy of increasing intelligence has already been developed in evolution as swarm intelligence. The collective intelligence of a termite population, for example, far exceeds the capabilities of individual animals. Only the co-operation of all animals creates the sophisticated termite structures. Insects use chemical codes to communicate with each other. Swarm intelligence is also pursued in robot technology.

Applied to humanity, the interaction of many people creates a collective human intelligence that is far superior to the individual [2]. This includes the intelligent infrastructures already mentioned. Increasingly, these collective systems are also making automated decisions. This does not require consciousness as we humans do.

(3) **New superintelligence:** An AI-system has new intellectual abilities which humans do not have.

This strategy is also inherent in the history of technology. Inventors and engineers find intelligent solutions to problems that were by no means predetermined in evolution. Robots that use the entire Internet as a memory and apply machine learning algorithms to big data at lightning speed have long been realisable.

The architecture of the human brain differs fundamentally from the digital technology of digital computers. In contrast to the targeted and conscious optimisation of technology, which took place in a short space of time, the brain architecture developed more or less randomly over millions of years of evolution under changing conditions and requirements. Biological nerve cells developed over long periods of time from cells that initially generated nerve signals incidentally, then increasingly, before finally specialising in the generation of action potentials for control and regulation tasks. This gave rise to highly sophisticated neurochemical signal processing with synapses and ion channels, which made our intellectual abilities possible in the first place.

On the other hand, biological neurons can only fire very slowly if you compare them with a modern microprocessor. This slowness was compensated for in the brain by an increased expansion of parallel signal processing and led to the enormous network density of the human brain. These complex brain networks and learning algorithms enabled the pattern recognition that is so crucial for the survival of animals, including humans. In contrast, signals in the von Neumann architecture of a computer are processed sequentially and separately in memory and processor units. The technology relies on the enormous speed of signal processing that is possible with silicon hardware according to Moore's law. On the other hand, this digital technology leads to the infamous 'von Neumann bottleneck' with enormous energy consumption and the resulting environmental impact.

Logically and mathematically, it can be proven that both approaches are equivalent — a (von Neumann) computer or a Turing machine and a (recurrent) neural network (with rational weights). In a superintelligence, the advantages of one approach could be used to compensate for the disadvantages of the other. For example, the material of technical microprocessors, transistors and memristors is more stable and resilient than biological neurons, axons and synapses. In the event of defects, they could be replaced like spent bulbs. Biological tissue, on the other hand, is subject to ageing processes and pathological changes (e.g. tumour formation), not all of which we have yet understood. Technical brain networks are therefore conceivable that could exchange signals much faster and more resiliently than biological neurons and synapses. This could be crucial for space travel, as organic tissue will be exposed to pathological changes caused by cosmic radiation, just like in humans. The advantages of human brains would have to be mapped in AI algorithms and realised on materials that are immune to the cosmic radiation damage of human tissue.

What scenarios are conceivable in which a superintelligence could develop? The taking off of a superintelligence is assumed to occur in several stages. The first stage would be a basic human level at which a human brain can be fully simulated by an individual AI system. This could be an artificial neuromorphic system. A computer that is logically and mathematically equivalent to this, as originally imagined by Turing, is by no means ruled out.

In principle, all living brains could be replaced by such an AI system. This would create a hybrid AI system that would be equivalent to the collective intelligence of humanity. On the way there, subsystems would have to emerge (e.g. intelligent infrastructures) that make increasingly independent decisions and set their own goals, because only then would they be able to determine consequences and outcomes better than humans. In this way, a threshold would be crossed that would ultimately lead to superintelligence.

Superintelligence is an AI system that is superior to individual and collective human intelligence (Fig. 1). An AI system with the ability to improve itself is referred to as 'seed AI'.

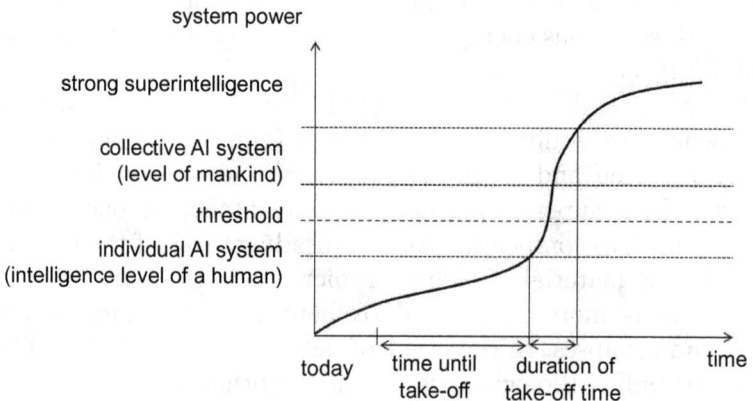

Fig. 1.   Take-off of superintelligence [1].

Nevertheless, a superintelligence also has the limits of

(a) logical–mathematical reasoning and proving with respect to computability, decidability and complexity
(b) physical laws.

The question remains as to whether this superintelligence will arrive at a specific point in time or whether it will develop in a continuous, long-term process. As early as 1965, the statistician I. J. Good, who had worked with Turing during the Second World War, predicted: 'The ultra-intelligent machine is therefore the last invention that man has to make' [3].

Technological singularity refers to the point in time when a superintelligence emerges. The mathematician V. Vinge published an article in 1993 entitled 'Technological Singularity', in which he linked the end of the era of humanity with technological singularity [4]. Computer scientist and author R. Kurzweil attributes the singularity to exponentially growing technologies [5]. This includes not only computing capacity according to Moore's Law, but also nano and sensor technology, as well as genetic engineering, neurotechnology and synthetic biology technologies to create new life forms.

What was dismissed years ago as scientifically dubious fantasies by science fiction authors is becoming increasingly realistic. In the end, however, the question remains as to how superintelligence can still be controlled as a service for humans after the singularity. In order to turn AI into a successful business model, influential robotics laboratories are also teaming up with military technology. A military arms race of AI weapon systems in space could initiate the development spiral towards a superintelligence that is no longer interested in the general welfare of humanity.

Decades ago, molecular biologists and genetic engineers warned against the misuse of their knowledge and expertise. In view of the increasing autonomy of AI systems, renowned scientists and technology entrepreneurs are now warning of a digital arms race in AI and drawing parallels with nuclear weapons. The unlimited energy of nuclear fission corresponds to the unlimited increase in intelligence of a superintelligence — in both cases uncontrollable.

How will innovation develop in cosmic civilisations?

Artificial intelligence will converge with other technological developments in order to enable the departure into space. The innovations required for this can be assessed in a scientific manner by asking the following questions:

- Which technologies are possible today?
- For which technologies do we already have the technical realisation possibilities to build them in the next few years (e.g. energy transport using laser beams, electric vehicles)?
- Which technologies are physically conceivable, but their technical realisation still fails due to many difficulties (e.g. fusion reactor, rocket propulsion through fusion)?
- Which technologies are physically conceivable, but without any currently foreseeable technical possibility of realisation (e.g. rocket propulsion using antimatter, vehicle propulsion using superconductors)?
- What role will nanotechnology and robotics play in space travel? How do the development stages of space technology depend on the development stages of human civilisation?

It is remarkable that the computer pioneer Konrad Zuse was already planning concrete projects for the future of AI in space in the 1960s. If, according to Zuse, the universe is a calculable cellular automaton, then automatons should also be used to colonise it. The theory of cellular automata began with the question of how automata should reproduce themselves in the same way as living organisms. The problem was solved mathematically by John von Neumann with a universal cellular automaton. However, the engineer and inventor Zuse was preoccupied with the technical problem of building a corresponding robot. At the beginning of the 1970s, Zuse started his project of self-reproducing systems with the construction of the SRS72 assembly line, which was to build a copy of itself with workpieces fed into it. The restored assembly line is now in the Deutsches Museum in Munich [6].

Zuse associated this with the vision of a technical germ cell that could reproduce itself with internal data storage and data processing by drawing on available raw materials in order to grow into a complex system like a biological organism. With such germ cells, according to Zuse, human civilisation could spread into space: Germ cells on one planet give rise to intelligent robot factories, which in turn produce germ cells that are shot to other planets in other star systems to repeat the self-reproduction process there. In 1980, American physicists described these scenarios as the 'von Neumann probe'. Unlike Zuse, however, von Neumann never mentioned such a space project.

New possibilities in materials research, such as those opened up by nanotechnology, will certainly be of central importance for technical self-reproduction. Initially, self-reproducing and more or less autonomous technologies will be integrated with humans in sociotechnical systems. The Internet of Things and Industry 4.0 are the first steps in this direction.

Cosmic innovation potential of information

Information and algorithms will be the universal categories with which not only technical but also the associated social, economic and societal changes can be recorded. The American astrophysicist Carl Sagan (1934–1996) therefore proposed a scale that measures civilisations according to the level of data processing [7]. His scale runs from the letters A to Z, each corresponding to increasing data capacities:

A Type A civilisation can only handle one million bits. This would be a level of development in which only spoken language can be used but not written language with documents. Think of primitive peoples, such as those discovered in the Amazon region. An ancient civilisation such as Greece, with its surviving written documents, has an estimated size of one billion bits and corresponds to a Type C civilisation on Sagan's scale. Sagan's assessment of the current civilisation was before the age of Big Data. With Big Data, we are on our way from the Peta (10^{15}) byte age to the Exa (10^{18}), Zetta (10^{21}) and Yotta (10^{24}) byte ages.

According to the ideas of Silicon Valley, things will only really get going after the singularity with superintelligence. By today's standards, the spread of an ever-changing human species in the universe will also require superintelligence.

Cosmic innovation potential of energy

The flip side of the amount of information required for this is an enormous demand for energy. Every type of civilisation depends on its energy consumption. The Russian astrophysicist Nicolai Kardashev had already considered in the 1960s how the level of development of future civilisations can be classified according to the possibilities of their energy consumption [8]. The result is a quantitative scale with measurable variables. Kardashev distinguished between three types of civilisation:

Type 1 civilisation dominates the energy of its planet. The consumable energy of a planet is determined by the fraction of incident light from its sun. With regard to the Earth, we can assume an estimated value of approximately 10^{17} watts. This is not just solar energy, which is now generated by solar power and photovoltaics. Fossil fuels are solar energy stored in dead plants. Wind, weather and ocean currents are also only made possible by solar energy. A civilisation of this type can master all these forms of energy. This currently seems utopian for humanity, but it is not physically impossible. Humanity is therefore still a Type 0 civilisation with energy consumption of less than 10^{17} watts.

In the mathematical theory of plasma physics, we have already put the fusion energy of the sun into formulae. However, the fusion reactor is still a long way off. According to Kardashev, this would be the first step towards a Type 2 civilisation: it would control the energy of the sun, i.e. approximately 10^{27} watts. This does not just mean solar cells, which are used to passively capture solar energy. The American physicist F. Dyson describes how such a civilisation constructs a gigantic sphere around its home star in order to absorb all of its radiation. A Type 3 civilisation is galactic and consumes the energy of billions of stars in the order of 10^{37} watts.

So far, we can only visualise the Kardashev scale in pictures as they are known from science fiction literature. The Type 1 civilisation would be the world of Flash Gordon, because all planetary energy sources can be used there. Even wind and weather are then completely controllable. The Type 2 civilisation is the planetary federation in *Star Trek*, which has already colonised a hundred nearby stars. Finally, the empire in the film *Star Wars* corresponds to a Type 3 civilisation: large parts of a galaxy with billions of stars are used. Under these conditions, superintelligence would spread throughout the universe.

Innovation and human rights in space

Back to the initial question of AI algorithms becoming independent in space travel: the behaviour of HALE in Kubrick's film will be conceivable in the near future. But not so much because the space robot is driven by existential fears and feelings like HALE, but because humans are overwhelmed, no longer have an overview of the complexity of a situation and the AI system has to intervene. Our often-hopeless discord on Earth, our selfishness and our inability to consider the overall well-being of this planet give us an idea of what could happen here in the future. However, it would be just as hopelessly naive to hope for a 'good' artificial intelligence that keeps human stupidity under control. Without the support of AI algorithms, it will not be possible to overcome the complex challenges of space travel. Without verification of these AI programmes, however, humanity is heading for an odyssey in space, of which Kubrick's vision was only a shadowy foreboding [9].

Just as human nature manifests itself on a daily basis, we will take conflicts and crises, brutality and barbarism with us into space. Since Archimedes and Leonardo da Vinci, innovation has been associated with ingenious and terrible war technology. Werner von Braun's American moon rocket Saturn V began with his construction of the retaliatory weapons V1 and V2 for the Nazi regime.

During the Cold War, rocket technology, which made it possible to fly to the moon, was the basis of attack and defence systems. The same applies today to the civilian and military use of satellite technology for global communication networks based on conventional communications technology or quantum technology. But people will also take their legal and standardisation systems with them into space. Initial approaches to transferring civil law with property regulations, contract law, etc. to space already exist.

What will basic rights look like in space? Humans will (have to) adapt to the living conditions in cosmic settlements. New findings in brain and cognitive research, medicine and psychology will change the way people see themselves, as will new technical possibilities offered by neuromorphic AI. Hybrid innovation potential in a symbiosis of humans and technology will determine everyday research.

What consequences follow from these insights into action and decision-making in complex systems beyond the mathematical models known today? All experiences show us that decision-making behaviour in political and economic systems is ultimately based on a deeper layer. People decide and act consciously or unconsciously on the basis of legal, cultural and religious values that have grown in different traditions around the world for centuries and have socialised and shaped them. We can therefore see these values as parameters for organising legal, cultural and religious dynamics. It is a global challenge to promote peaceful coexistence and cultural balance in order to prevent the crash of civilisation in its complex non-linear dynamics.

> From the point of view of non-linear dynamics, it is about creating common 'order parameters' to ensure the global governance of this planet, minimise conflicts and reduce complexity. We need to trigger suitable impulses and signals in education and training so that this integration can grow and develop. It cannot be prescribed or programmed. This insight is also conveyed by complexity and innovation research.

References

[1] Bostrom, N. (2014), *Superintelligence. Scenarios of a Coming Revolution*, Berlin. Oxford University Press: Oxford.

[2] Shanahan, M. (2010), *Embodiment and the Inner Life. Cognition and Consciousness in the Space of Possible Minds*, New York. Imperial College Press: London.

[3] Good, I. J. (1965), Speculations concerning the first ultraintelligent machine, in Alt, F. L. and Robinoff, M. (Eds.) *Advances in Computers*, New York, pp. 31–88. Academic Press.

[4] Vinge, V. (1993), The coming technological singularity: How to survive in the post-human era, *Vision-21: Interdisciplinary Science and Engineering in the Era of Cyberspace. NASA Conference Publication*, 10(129), pp. 11–22. NASA Lewis Research Center.

[5] Kurzweil, R. (2005), *The Singularity Is Near. When Humans Transcend Biology*, New York. Pinguin Publishing Group.

[6] Eibisch, N. (2011), Eine Maschine baut eine Maschine baut eine Maschine ... , *Kultur und Technik,* 1, pp. 48–51.

[7] Slovskii, I. S. and Sagan, C. (1966), *Intelligent Life in the Universe,* Holden-Day: San Francisco.

[8] Kardashev, N. S. (1964), Transmission of information by extraterrestrial civilizations, *Soviet Astronomy,* 8(2), pp. 217–221.

[9] Mainzer, K. (2022), Künstliche Intelligenz auf interstellarer Mission. Robonauten, Androiden, Cyborgs: Raumfahrer der Zukunft?, in Zaun, H. (Ed.) *Expedition ins Sternenmeer. Perspektiven, Chancen und Risiken interstellarer Raumfahrt,* Springer: Berlin, pp. 221–236.

Index

www.ingramcontent.com/pod-product-compliance
Lightning Source LLC
Chambersburg PA
CBHW050640190326
41458CB00008B/2355